Lecture Notes
in Control and Information Sciences   256

Editor: M. Thoma

Springer
London
Berlin
Heidelberg
New York
Barcelona
Hong Kong
Milan
Paris
Singapore
Tokyo

Eva Zerz

Topics in Multidimensional Linear Systems Theory

With 4 Figures

Springer

Series Advisory Board

Author

Eva Zerz, PhD
Department of Mathematics, University of Kaiserslautern, 67663 Kaiserslautern, Germany

ISBN 1-85233-336-7 Springer-Verlag London Berlin Heidelberg

British Library Cataloguing in Publication Data
Zerz, Eva
   Topics in multidimensional linear systems theory. -
   (Lecture notes in control and information sciences ; 256)
   1.Linear systems 2.Control theory
   I.Title
   629.8'32
   ISBN 1852333367

Library of Congress Cataloging-in-Publication Data
Zerz, Eva.
   Topics in multidimensional linear systems theory / Eva Zerz.
       p. cm. -- (Lecture notes in control and information sciences, ISSN 0170-8643 ; 256)
   Includes bibliographical references and index.
   ISBN 1-85233-336-7 (alk. paper)
   1. Linear systems. I. Title. II. Series.
   QA402.Z47 2000
   003'.74--dc21   00-041916

Typesetting: Camera ready by author

69/3830-543210 Printed on acid-free paper SPIN 10770364

Preface

The first contributions to mathematical systems and control theory date back as far as J. C. Maxwell's 1868 paper "On Governors." Still, it took another hundred years for systems and control to be recognized as a mathematical discipline in its own right. This is especially due to the pioneering work of R. E. Kalman.

In the 1980s, J. C. Willems proposed an alternative approach to systems and control, based on studying the solution spaces rather than the equations that generate them, and on replacing the causal input–output or input–state–output framework by a setting in which all signals connected with a system are treated equally. This reflects the fact that in complex systems, it may be hard or even impossible to decide what is cause and what is effect. Willems' behavioral approach postulates that the distinction between system variables that are free and system variables that are determined by others should be derived from the model, in contrast to being imposed on it *a priori*. Moreover, one interpretation may be just as legitimate as another, *i.e.*, there may be several admissible input–output interpretations for one and the same system.

Dynamic systems evolve in time, that is, they depend on one free parameter. The world, however, is not one-dimensional, and very soon there were attempts to develop a multidimensional generalization of systems theoretic concepts. Mathematically, the transition from one-dimensional to multidimensional systems corresponds to that from ordinary to partial differential or difference equations. The two-dimensional case, with its applications in image processing, is particularly interesting, and it has its own distinctive features. Local state space models for two-dimensional systems have been proposed by R. P. Roesser, S. Attasi, E. Fornasini and G. Marchesini, T. Kaczorek and others (this and the following lists of authors make no claim whatsoever to completeness).

The behavioral approach has proved to be particularly fruitful for systems in more than one independent variable, and two-dimensional behaviors have been studied by E. Fornasini, P. Rocha, M. E. Valcher, J. C. Willems, and S. Zampieri. For arbitrary multidimensional systems, the polynomial matrix approach has been initiated by N. K. Bose and D. C. Youla. The profound work of U. Oberst provides an algebraic framework for the study of multidimensional behaviors.

Organization and Summary

In Chapter 1, we study the concept of controllability, which is the foundation of control theory. The classical notion is that by choosing a suitable input, it is possible to pass from any given initial state to any desired final state, in finite time. This notion seems to be very much related with input–state concepts, but Willems showed that controllability is a built-in property of a system behavior that does, for instance, not depend on the distinction between input and states, a fact that is very far from engineering intuition. Even from classical results like the Hautus rank condition for controllability, it becomes apparent that the mathematical key to controllability is primeness of polynomial matrices. Subtle distinctions have to be made in the multivariate setting.

The primeness notion developed in the first chapter is applied to the problem of co-prime factorization of rational matrices in Chapter 2, and the kernel (image) representations associated to left (right) factorizations are discussed. The interpretation of multivariate rational matrices as transfer functions of multidimensional systems is not as obvious though as in the one-dimensional case, as the input–output relations are typically *not* given by a well-defined convolution. This is due to the fact that causality, a perfectly reasonable assumption in the one-dimensional setting ("future input values cannot influence present output values"), becomes a rather unnatural requirement with systems that depend on several independent variables.

As is well-known in the case of one-dimensional systems, the co-prime factorization approach provides a framework for the solution of a variety of control problems, for instance, controller synthesis by means of the Youla parameterization of all stabilizing controllers of a plant. In Chapter 3, we consider these questions in the multidimensional setting, *i.e.*, we study co-primeness and factorization concepts over rings of stable transfer functions, and we apply the results to the problem of stabilizing a plant (represented by its multivariate rational transfer matrix) by dynamic output feedback.

Chapter 4 defines a class of admissible systems of linear partial difference or differential equations for which a transfer function is well-defined. We study several equivalence relations for this system class, including a generalization of H. H. Rosenbrock's classical concept of strict system equivalence, which can be characterized in terms of certain isomorphisms between the solution spaces. Several results by P. Fuhrmann and L. Pernebo have nice generalizations to the multidimensional setting.

First order representations of multidimensional behaviors are studied in Chapter 5. The classical local state-space systems, like the Roesser or the Fornasini–Marchesini models, are only capable of representing causal transfer functions. Dropping the causality requirement, we consider general first order representations which involve manifest system variables (the ones we are interested in, e.g., "inputs and outputs") and latent variables (generalized "states" that occur in the modeling process, as well as with the reduction to first order). This leads naturally to so-called output–nulling representations, consisting of a first order dynamic and a static relation between the manifest and latent variables. These representations have been studied, in the one-dimensional setting, by J. C. Willems, M. Kuijper, and S. Weiland. Deriving output–nulling from kernel representations, as well as the dual construction of driving–variable from image representations, is based on the representation of a polynomial matrix in terms of a linear fractional transformation.

These linear fractional transformations are studied in more detail in Chapter 6, where a construction technique is given that can be interpreted as a realization algorithm for multidimensional finite–impulse–response filters. The size of such a linear fractional representation can be reduced by a transformation to trim form. This reduction does not change the value of the linear fractional transformation, and thus can be seen as an exact model reduction method. We also describe an approximative model reduction that is based on an appropriate notion of balancing for a class of linear fractional representations. Linear fractional transformations are prominent in robust control, and we point out some of the connections. In particular, we establish a result on the stability radius of descriptor systems, with considerably relaxed assumptions compared to previously published results. As an example, an upper bound for the stability radius of an electrical circuit is computed.

The final chapter collects some known facts from the mathematical theory of networks and circuits, and thus sheds some light on a phenomenon observed in the example of the previous chapter.

Acknowledgements

I am grateful to Prof. Dr. Ulrich Oberst, who introduced me to the subject of multidimensional systems, and whose ground-breaking work has fundamental importance for everything I have done in this area. Further inspiration came from Prof. Dr. Jean-François Pommaret and Prof. Dr. Maria Elena Valcher. I thank Dr. Jeffrey Wood (co-author of Section 1.4 of this work) for fruitful discussions and cooperation. Special thanks go to Prof. Dr. Nirmal K. Bose, Prof. Dr. Paula Rocha, and Prof. Dr. Dieter Prätzel-Wolters.

Kaiserslautern, June 2000   *Eva Zerz*

Table of Contents

List of Acronyms

BIBO	bounded input, bounded output
DV	driving variable
FLP	factor left prime
FRP	factor right prime
FSSE	strict system equivalence in the Fuhrmann sense
GCLF	greatest common left factor
GCRF	greatest common right factor
GFLP	generalized factor left prime
GFRP	generalized factor right prime
IAM	indefinite admittance matrix
IO	input–output
ISO	input–state–output
KCL	Kirchhoff's current law
KVL	Kirchhoff's voltage law
LFT	linear fractional transformation
LMI	linear matrix inequality
MLA	minimal left annihilator
MLP	minor left prime
MNA	modified nodal analysis
MRA	minimal right annihilator
MRP	minor right prime
ON	output–nulling
PDE	partial difference/differential equation
RCL	resistance, capacitance, inductance
RSSE	strict system equivalence in the Rosenbrock sense
SC	strongly controllable
SSE	strict system equivalence
WLP	weakly zero left prime
WRP	weakly zero right prime
ZRP	zero right prime
ZLP	zero left prime

1. Controllability of Multidimensional Systems

Primeness and co-primeness concepts play an important role in the polynomial matrix approach to systems and control theory, which was initiated by Rosenbrock [63]. The notions of observability and controllability lie at the very heart of systems theory, and they are closely related to questions of primeness.

In the behavioral approach to systems theory, introduced by Willems [79, 80, 81], a dynamical system is a triple $\Sigma = (T, W, \mathcal{B})$, where T is the time domain, W the signal space, and $\mathcal{B} \subset W^T$ the behavior, that is, a family of trajectories evolving over T, taking their values in W, and satisfying the laws which govern the system. We will mainly consider the discrete time domain $T = \mathbb{N}$ (or $T = \mathbb{N}^r$ in the multidimensional case) and the signal space $W = \mathbb{R}^q$, where $\mathbb{N}$ and $\mathbb{R}$ denote the set of non-negative integers and the field of real numbers, respectively. Once the time domain and the signal space are fixed, systems and their behaviors may be identified.

A discrete behavior that is linear, shift–invariant, and complete, admits an autoregressive representation. Thus, it can be written as the kernel of a difference operator which is represented by a polynomial matrix in which the indeterminate has been replaced by the left shift difference operator σ, defined by $(\sigma w)(i) = w(i+1)$ for $w \in W^T$. Thus the behavior $\mathcal{B} = \ker(R(\sigma))$, or $\mathcal{B} = \ker(R)$ for short, is the solution space of a linear system of difference equations with constant coefficients.

In this framework, controllability can be defined as an intrinsic system property that does not depend on the choice of a particular representation. $\mathcal{B}$ is called controllable if it admits a kernel representation with a left prime polynomial matrix R. Controllable behaviors have image representations, that is, they can be written as $\mathcal{B} = \operatorname{im}(M(\sigma)) = \operatorname{im}(M)$, where M is a polynomial matrix, i.e., $M(\sigma)$ is another linear constant-coefficient difference operator.

In a partition $R_1(\sigma)w_1 = R_2(\sigma)w_2$ of the system laws $R(\sigma)w = 0$, w_2 is said to be observable from w_1 if knowledge of w_1 yields knowledge of w_2, which reduces to the requirement that $w_1 = 0$ implies $w_2 = 0$ because of linearity. This can be shown to be equivalent to right primeness of R_2.

Rocha [59] generalized this notion of observability to 2D behaviors over $T = \mathbb{Z}^2$. In a kernel representation of such a behavior, the entries of the representing matrix are bivariate Laurent polynomials in which the two indeterminates are identified with the invertible left and down shift difference operators σ_1 and σ_2, given by $(\sigma_1 w)(i,j) = w(i+1,j)$ and $(\sigma_2 w)(i,j) = w(i,j+1)$. Hence, a 2D behavior $\mathcal{B} = \ker(R(\sigma_1, \sigma_2, \sigma_1^{-1}, \sigma_2^{-1})) = \ker(R)$ is the solution set of a linear system of partial difference equations.

Generalizing from the one-dimensional case, in a partition $w = [w_1^T, w_2^T]^T$ of the system variables with a corresponding decomposition $R_1 w_1 = R_2 w_2$ of the system laws, w_2 is called strongly observable from w_1 if it is uniquely determined by w_1, which is equivalent to *zero* right primeness of R_2. A weaker notion of observability is obtained in the case of a *factor* right prime matrix R_2, where w_1 determines w_2 up to a finite-dimensional real vector space.

Rocha and Willems [61] also introduced a notion of controllability for 2D systems in the behavioral framework, which was shown to hold for 2D behaviors that admit a kernel representation $\mathcal{B} = \ker(R)$ with a factor left prime matrix R. If the representing matrix R is even zero left prime, the behavior is called strongly controllable [60].

Various primeness concepts for multivariate polynomial matrices were also treated by Bose [3], Fornasini and Valcher [24], Youla and Gnavi [88] and, in the 2D case, Morf *et al.* [46] and Kung *et al.* [38].

Summing up, primeness of polynomial matrices is the mathematical key to controllability of linear systems. This is why we start, after presenting some preliminary material in Section 1.1, with a discussion of multivariate polynomial matrices and their primeness in Section 1.2. Algorithms for testing primeness properties are given; they are based on computer algebraic techniques, mainly, the theory of Gröbner bases. It is crucial to note that various characterizations of primeness of polynomial matrices that are mutually equivalent in the one-dimensional case of univariate polynomials, may lead to different primeness concepts in the multivariate situation. This fact will give rise to a more refined view to observability and controllability, and controllable–autonomous decompositions of multidimensional systems in Section 1.3. After intuitively giving the definition of controllability that seems to be most appropriate for multidimensional systems, it is then shown in Section 1.4 that this notion has an interpretation in terms of concatenability of trajectories, just as should be expected from a behavioral controllability notion. The question of existence of image representations is addressed in a slightly different setting in Section 1.5, where we discuss the "parameterizability" of linear constant-coefficient partial differential equations.

1.1 Preliminaries and Notation

1.1.1 Gröbner Bases

Let $\mathcal{D} = \mathbb{R}[z] = \mathbb{R}[z_1, \ldots, z_r]$ denote the polynomial ring over $\mathbb{R}$ in r indeterminates z_i. A **Gröbner basis** is a special set of generators of a non-zero ideal $\mathcal{I} \subseteq \mathcal{D}$. Buchberger's algorithm [7] for computing Gröbner bases can be interpreted as a generalization of the Euclidean algorithm to the multivariate situation, and it is implemented in many customary computer algebra systems.

As a **term order** [2, p. 189] on the set of monic monomials or terms

$$T = \{z^n = z_1^{n_1} \cdots z_r^{n_r}, \; n = (n_1, \ldots, n_r) \in \mathbb{N}^r\},$$

we will always consider (reverse–) **lexicographic** orders, where for some permutation i of the indeterminates, $z^n < z^m$ if there is a $j \in \{1, \ldots, r\}$ such that $n_{i(k)} = m_{i(k)}$ for $k > j$ and $n_{i(j)} < m_{i(j)}$. In particular, $z_{i(1)} < z_{i(2)} < \cdots < z_{i(r)}$, whence the indeterminates $z_{i(1)}$ and $z_{i(r)}$ are called the least and the greatest variable with respect to this order, respectively. Since any polynomial $h \in \mathcal{D}$ admits a unique representation $h = \sum_{t \in T} h(t)t$ with $h(t)$ the real coefficients of h, a term order makes it possible to define the **head term (ht)** of $h \neq 0$,

$$\mathrm{ht}(h) = \max \mathrm{supp}(h),$$

where $\mathrm{supp}(h) = \{t \in T, h(t) \neq 0\}$ is a finite set.

A finite subset $G \subset \mathcal{I} \setminus \{0\}$ is called a Gröbner basis of $\mathcal{I}$ if for any non-zero $h \in \mathcal{I}$ there exists a $g \in G$ such that $\mathrm{ht}(g)$ divides $\mathrm{ht}(h)$. There is no loss of generality in assuming that the Gröbner bases considered here are always reduced [2, p. 209]. Once a term order is fixed, the reduced Gröbner basis of a non-zero ideal is uniquely determined, and two ideals are equal iff their reduced Gröbner bases are equal.

The theory can easily be generalized to polynomial modules. We will use

$$z^n e_i < z^m e_j \iff (z^n < z^m \text{ lexicographically}) \text{ or } (n = m \text{ and } i < j) \quad (1.1)$$

as a term order on the set $\{z^n e_i, \; n \in \mathbb{N}^r, \; 1 \leq i \leq l\}$, where e_i denotes the i-th standard basis vector. For a polynomial matrix $R \in \mathcal{D}^{g \times q}$, let $cG(R)$ denote the column Gröbner matrix of R, i.e., a matrix whose columns are a reduced Gröbner basis of the column module $\mathrm{CM}(R) = R\mathcal{D}^q \subseteq \mathcal{D}^g$. The column Gröbner matrix is unique up to a permutation of its columns. It allows one to decide equality of polynomial column modules, since for a fixed term order, two matrices generate the same column module iff their column Gröbner matrices are equal. When speaking of equality of Gröbner matrices, this is meant to signify "equality up to permutation of columns." The extended Gröbner basis algorithm also computes the transition matrices X and Y with $R = cG(R)X$ and $cG(R) = RY$.

Analogously, the row Gröbner matrix $rG(R)$ is defined as a matrix whose rows are a reduced Gröbner basis of the row module $\mathrm{RM}(R) = \mathcal{D}^{1 \times g} R \subseteq \mathcal{D}^{1 \times q}$.

1.1.2 Elementary Algebraic Geometry

An **algebraic set** is the set of common zeros of a finite set of polynomials. By Hilbert's basis theorem [39, p. 10], any polynomial ideal $\mathcal{I} \subseteq \mathcal{D} =$

$\mathbb{R}[z_1, \ldots, z_r]$ has a finite system of generators. Thus

$$\mathcal{V}(\mathcal{I}) = \{\xi \in \mathbb{C}^r, \ f(\xi) = 0 \text{ for all } f \in \mathcal{I}\}$$

is an algebraic set, called the **variety** of $\mathcal{I}$. Conversely, an ideal

$$\mathcal{J}(V) = \{f \in \mathcal{D}, \ f(\xi) = 0 \text{ for all } \xi \in V\}$$

is associated to any algebraic set V. It is easy to see that $\mathcal{V}(\cdot)$ and $\mathcal{J}(\cdot)$ are order-reversing mappings. Hilbert's Nullstellensatz [39, p. 18], which holds over any algebraically closed field (here: $\mathbb{C}$), says that

$$\mathcal{J}(\mathcal{V}(\mathcal{I})) = \text{Rad}(\mathcal{I}) = \{f \in \mathcal{D}, \ \exists l \in \mathbb{N} : \ f^l \in \mathcal{I}\}.$$

The following weaker form of the Nullstellensatz [39, p. 16] is also useful: $\mathcal{V}(\mathcal{I}) = \emptyset$ iff $1 \in \mathcal{I}$, that is, $\mathcal{I} = \mathcal{D}$.

The **dimension** of a non-empty algebraic set V is the maximal dimension of its irreducible components. By convention, $\dim(\emptyset) = -1$. A non-constant polynomial $h \in \mathcal{D}$ defines the algebraic hyper-surface $\mathcal{V}(h) = \{\xi \in \mathbb{C}^r, \ h(\xi) = 0\}$ of dimension $r - 1$.

The dimension of a proper polynomial ideal $\mathcal{I} \subset \mathcal{D}$ is defined to be the Krull dimension [39, p. 40] of the ring $\mathcal{D}/\mathcal{I}$, and it is denoted by $\dim(\mathcal{I})$. It can also be characterized as follows [2, p. 270]: The dimension of $\mathcal{I}$ is the maximal number k such that there exist indeterminates $z_{i(1)}, \ldots, z_{i(k)}$ with

$$\mathcal{I} \cap \mathbb{R}[z_{i(1)}, \ldots, z_{i(k)}] = \{0\}. \tag{1.2}$$

This condition can be checked using a Gröbner basis technique [2, p. 257]: Choose a lexicographic term order with $z_{i(1)} < \cdots < z_{i(k)} < \cdots < z_{i(r)}$ and compute a Gröbner basis G of $\mathcal{I}$. Then $G \cap \mathbb{R}[z_{i(1)}, \ldots, z_{i(k)}]$ is a Gröbner basis of $\mathcal{I} \cap \mathbb{R}[z_{i(1)}, \ldots, z_{i(k)}]$.

We set $\dim(\mathcal{D}) = -1$. A fundamental fact of algebraic geometry [39, p. 40] is that the dimension of an algebraic set V coincides with the dimension of its coordinate ring $\mathcal{D}/\mathcal{J}(V)$, and thus with $\dim(\mathcal{J}(V))$. Furthermore, $\dim(\mathcal{I}) = \dim(\text{Rad}(\mathcal{I}))$, hence

$$\dim(\mathcal{V}(\mathcal{I})) = \dim(\mathcal{I}). \tag{1.3}$$

1.2 Primeness of Multivariate Polynomial Matrices

Right primeness properties hold whenever the corresponding left primeness assertions are true for the transposed matrix. We will therefore concentrate on the discussion of left primeness notions.

Let $\mathcal{D} = \mathbb{R}[z] = \mathbb{R}[z_1, \ldots, z_r]$ denote the polynomial ring over $\mathbb{R}$ in r indeterminates z_i. Its quotient field, the field of rational functions, is denoted by $\mathcal{K} = \mathbb{R}(z) = \mathbb{R}(z_1, \ldots, z_r)$. Let R be a matrix with entries in $\mathcal{D}$, $R \in \mathcal{D}^{g \times q}$, and $g \leq q$. Let $m_1, \ldots, m_k \in \mathcal{D}$ be the $g \times g$ minors of R. The ideal generated by these minors is denoted by $\mathcal{I} = \langle m_1, \ldots, m_k \rangle = \sum_i \mathcal{D} m_i \subseteq \mathcal{D}$. In the following, we will simply write "minor," when referring to these full order sub-determinants of R.

1.2.1 Zero and Minor Primeness

Definition 1 *The matrix R is called*

1. **zero left prime (ZLP)** *if its minors possess no common zero in $\mathbb{C}^r$,*
2. **weakly zero left prime (WLP)** *if the minors have only a finite number of common zeros in $\mathbb{C}^r$, and*
3. **minor left prime (MLP)** *if the minors are devoid of a non-trivial common factor.*

If the rank of R as a matrix over the quotient field of rational functions is less than g, then all the m_i are identically zero, hence R does not satisfy any of the above primeness conditions. We exclude this case for the time being. For a polynomial matrix R of full row rank define the algebraic variety of rank singularities

$$
\begin{aligned}
\mathcal{V}(R) &= \{\xi \in \mathbb{C}^r, \, \mathrm{rank}\,(R(\xi)) < \mathrm{rank}\,(R)\} \\
&= \{\xi \in \mathbb{C}^r, \, m_1(\xi) = \ldots = m_k(\xi) = 0\} = \mathcal{V}(\mathcal{I}).
\end{aligned}
$$

Since R is assumed to have full row rank, we have $\dim \mathcal{V}(R) \leq r - 1$.

Lemma 1.2.1. *The matrix R is*

1. *ZLP iff $\mathcal{V}(R)$ is empty ($\dim \mathcal{V}(R) = -1$),*
2. *WLP iff $\mathcal{V}(R)$ is finite ($\dim \mathcal{V}(R) \leq 0$), and*
3. *MLP iff $\mathcal{V}(R)$ contains no algebraic hyper-surface ($\dim \mathcal{V}(R) \leq r - 2$).*

Proof. The first two assertions are obvious from the definitions. We show that the existence of a non-constant common factor of the minors m_i is equivalent to $\dim \mathcal{V}(R) = r - 1$. Indeed, if the minors m_i share a non-constant common factor, say h, we have

$$\mathcal{V}(h) \subseteq \mathcal{V}(R) \tag{1.4}$$

and hence $r - 1 = \dim \mathcal{V}(h) \leq \dim \mathcal{V}(R)$. The case that $\dim \mathcal{V}(R) = r$ is excluded by the rank condition on R. Conversely, if $\dim \mathcal{V}(R) = r - 1$, $\mathcal{V}(R)$ contains an irreducible hyper-surface, that is, some $\mathcal{V}(h)$, where h is an irreducible polynomial. In particular, the principal ideal $\langle h \rangle$ is perfect, *i.e.*, $\mathrm{Rad}(\langle h \rangle) = \langle h \rangle$. But (1.4) implies, by Hilbert's Nullstellensatz,

$$\langle h \rangle = \mathrm{Rad}(\langle h \rangle) = \mathcal{J}(\mathcal{V}(h)) \supseteq \mathcal{J}(\mathcal{V}(R)) = \mathcal{J}(\mathcal{V}(\mathcal{I})) = \mathrm{Rad}(\mathcal{I}) \supseteq \mathcal{I}.$$

Thus any element of $\mathcal{I}$, in particular each minor m_i, is contained in $\langle h \rangle$, that is, it must be a multiple of h. $\square$

From the lemma, it is easy to see that in the univariate case ($r = 1$), ZLP and MLP are equivalent concepts, whereas WLP is always true. In the bivariate situation ($r = 2$), ZLP is stronger than WLP, and WLP is equivalent to MLP. For $r \geq 3$, the three primeness definitions are mutually inequivalent, but always

$$ZLP \quad \Rightarrow \quad WLP \quad \Rightarrow \quad MLP.$$

Now the following corollary is an immediate consequence of (1.3).

Corollary 1 *The matrix R is*

1. *ZLP iff* $\dim(\mathcal{I}) = -1$, *that is,* $\mathcal{I} = \mathcal{D}$, *or*

$$\mathcal{I} \ni 1,$$

2. *WLP iff $\mathcal{I}$ is at most zero-dimensional, i.e., iff $\mathcal{D}/\mathcal{I}$ is a finite-dimensional real vector space, or equivalently,*

$$\mathcal{I} \ni d_i \in \mathbb{R}[z_i] \setminus \{0\} \ for \ 1 \leq i \leq r,$$

3. *MLP iff there is no proper principal ideal that contains $\mathcal{I}$, i.e., iff the dimension of $\mathcal{I}$ is less than $r - 1$ or*

$$\mathcal{I} \ni e_i \in \mathbb{R}[z_1, \ldots, z_{i-1}, z_{i+1}, \ldots z_r] \setminus \{0\} \ for \ 1 \leq i \leq r.$$

In fact, it seems to be conceptually clearer to use the characterization of Corollary 1, which is in terms of the dimension of ideals rather than varieties, for the definition of ZLP, WLP and MLP. This allows to generalize these primeness notions to arbitrary coefficient fields (here: $\mathbb{R}$) without having to deal with their algebraic closures.

The conditions of Corollary 1 can be checked via the Gröbner basis algorithm as follows: $\mathcal{I} = \mathcal{D}$ iff the reduced Gröbner basis of $\mathcal{I}$ with respect to any term order is $\{1\}$. $\mathcal{I}$ is at most zero-dimensional iff for all i, the Gröbner basis of $\mathcal{I}$ with respect to a lexicographic term order with z_i the least variable, contains an element $d_i \in \mathbb{R}[z_i]$. The dimension of $\mathcal{I}$ is less than $r - 1$ iff for all i, the Gröbner basis of $\mathcal{I}$ with respect to a lexicographic term order with z_i the greatest variable, contains an element $e_i \in \mathbb{R}[z_1, \ldots, z_{i-1}, z_{i+1}, \ldots, z_r] =: \mathbb{R}[z \setminus z_i]$.

Instead of dealing with the ideal generated by the minors of R, it is sometimes more desirable to directly investigate the column module of R.

Lemma 1.2.2. *The matrix R is ZLP iff there is a matrix $X \in \mathcal{D}^{q \times g}$ that solves the Bézout equation*

$$RX = I_g,$$

that is, iff $\mathrm{CM}(R) = \mathcal{D}^g$. *It is WLP iff for $1 \leq i \leq r$, there are matrices $X_i \in \mathcal{D}^{q \times g}$ such that*

$$RX_i = d_i I_g, \quad 0 \neq d_i \in \mathbb{R}[z_i],$$

that is, iff $\mathcal{D}^g / \mathrm{CM}(R)$ is a finite-dimensional real vector space, and R is MLP iff for $1 \leq i \leq r$ there are matrices $X_i \in \mathcal{D}^{g \times q}$ such that

$$RX_i = e_i I_g, \quad 0 \neq e_i \in \mathbb{R}[z \setminus z_i].$$

In terms of Gröbner matrices, R is ZLP iff $\mathrm{cG}(R) = I_g$, WLP iff for all i, the column Gröbner matrix with respect to a lexicographic order with z_i the least variable, contains a non-singular $g \times g$ sub-matrix depending only on z_i, and MLP iff for all i, the column Gröbner matrix with respect to a lexicographic order with z_i the greatest variable, contains a non-singular $g \times g$ sub-matrix independent of z_i.

Proof. The conditions for zero and minor primeness were derived in [88]. Sufficiency follows from the Cauchy-Binet theorem [27, p. 27], which says that $\det(RX) = \sum m_j p_j$, where m_j and p_j are corresponding minors of R and X, respectively. The main idea for the proof of necessity consists in the construction of polynomial matrices Z_j that isolate each individual minor m_j of R in the form $RZ_j = m_j I_g$. The construction in [88] is a bit cumbersome though. In fact, it suffices to note that for every minor m_j, there exists a permutation matrix Π_j such that

$$R\Pi_j = \left[\begin{array}{cc} R_j^{(1)} & R_j^{(2)} \end{array} \right]$$

with a square matrix $R_j^{(1)}$ and $\det(R_j^{(1)}) = m_j$. Let $\mathrm{adj}\,(X)$ denote the adjoint of a square matrix X, that is, $X\,\mathrm{adj}\,(X) = \det(X)I$. Then

$$RZ_j := R\Pi_j \left[\begin{array}{c} \mathrm{adj}\,(R_j^{(1)}) \\ 0 \end{array} \right] = m_j I_g.$$

If R is ZLP, we have $1 \in \mathcal{I}$, hence there exist polynomials p_j such that $\sum m_j p_j = 1$. But then $\sum RZ_j p_j = I_g$. The conditions for weak zero and minor primeness can be proven analogously. □

For the characterization in terms of Gröbner matrices, the following fact is crucial (compare with the remark following (1.2)): If $\mathrm{cG}(R)$ is the Gröbner matrix of $\mathrm{CM}(R)$ with respect to a lexicographic order $z_{i(1)} < z_{i(2)} < \dots < z_{i(r)}$, then the sub-matrix of $\mathrm{cG}(R)$ consisting of all columns that depend only on $z_{i(1)}, \dots, z_{i(k)}$ is a Gröbner matrix of $\mathrm{CM}(R) \cap \mathbb{R}[z_{i(1)}, \dots, z_{i(k)}]^g$. Note that this generalization to the module case depends on our particular choice of the term order (1.1).

1.2.2 Factor Primeness

Definition 2 *A full row rank matrix R is* **factor left prime (FLP)** *if any square left factor is unimodular, that is, if the existence of a factorization $R = DR_1$ with a square matrix D implies that $\det(D)$ is a non-zero constant.*

A factorization $R = DR_1$ with D square clearly implies that the determinant of D divides all minors of R. Hence R is FLP if it is MLP. The converse, however, is only true for $r \leq 2$. Also, in contrast to zero and minor primeness, factor primeness cannot be decided by means of the variety $\mathcal{V}(R)$ alone. This fact is illustrated by the following example.

Example 1.2.1. The non-zero minors of both the matrices

$$R = \begin{bmatrix} z_1 & 0 & -z_2 \\ 0 & z_1 & z_3 \end{bmatrix} \quad \text{and} \quad \tilde{R} = \begin{bmatrix} z_1 & 0 & -z_2 & -z_3 \\ 0 & z_1 & 0 & 0 \end{bmatrix}$$

are $z_1^2, z_1 z_2$, and $z_1 z_3$, hence $\mathcal{V}(R) = \mathcal{V}(\tilde{R})$. But $\tilde{R}$ admits a factorization with the left factor $\mathrm{diag}(1, z_1)$ of determinant z_1, whereas R is FLP. Any factorization $R = DR_1$ requires $R_1 \zeta = 0$, where $\zeta = (z_2, -z_3, z_1)^T$. This implies that all entries of R_1 are polynomials whose constant coefficients are zero, that is, they are contained in the maximal ideal $\mathcal{J} = \langle z_1, z_2, z_3 \rangle$. But then the minors of R_1 are elements of $\mathcal{J}^2$, from which it is evident that none of the minors can have a linear head term. This excludes the case $\det(D) = z_1$. Thus $\det(D)$ has to be constant.

For a systems theoretic interpretation of factor primeness, the case that R is not of full row rank may no longer be neglected for polynomial rings in more than two indeterminates ($r > 2$), consider e.g., the example given below. More precisely, the systems that admit a full row rank representation are those whose **projective dimension** is at most one [48]. Consider the finitely generated module $\mathcal{M} = \mathrm{cok}(R) = \mathcal{D}^{1 \times q}/\mathcal{D}^{1 \times g}R$, where $\mathcal{D} = \mathbb{R}[z]$. For any $k \geq 1$, there is an exact sequence

$$0 \to M_k \to \ldots \to M_1 \to M_0 \to M_{-1} = \mathcal{M} \to 0 \tag{1.5}$$

with free modules $M_0, \ldots, M_{k-1}$. Such sequences can be constructed as follows: For $k = 1$, we have the trivial exact sequence

$$0 \longrightarrow \mathcal{D}^{1 \times g}R \hookrightarrow \mathcal{D}^{1 \times q} \to \mathcal{M} \to 0.$$

Combining this with the exact sequence

$$0 \to \ker(R) \hookrightarrow \mathcal{D}^{1 \times g} \xrightarrow{R} \mathcal{D}^{1 \times g}R \to 0, \tag{1.6}$$

we construct a sequence with $k = 2$,

$$0 \to \ker(R) \hookrightarrow \mathcal{D}^{1 \times g} \xrightarrow{R} \mathcal{D}^{1 \times q} \to \mathcal{M} \to 0. \tag{1.7}$$

Next, a **minimal left annihilator (MLA)** of R is computed by means of a Gröbner basis technique (see, for instance, [2, Section 10.5]). The algorithm has been implemented in various modern computer algebra systems, e.g., SINGULAR [28]. A $\mathcal{D}$-matrix Q is an MLA of R iff

1. Q is a left annihilator of R, that is, $QR = 0$, and
2. any left annihilator of R is a multiple of Q, that is, $Q_1 R = 0$ implies $Q_1 = XQ$ for some $\mathcal{D}$-matrix X.

In other words: The module $\ker(R) = \{\xi \in \mathcal{D}^{1 \times g}, \xi R = 0\}$ is generated by the rows of $Q \in \mathcal{D}^{m \times g}$, that is,

$$\ker(R) = \mathcal{D}^{1 \times m} Q.$$

In particular, $\operatorname{rank}(Q) + \operatorname{rank}(R) = g$. But now we have an exact sequence like (1.6) for Q instead of R, and we can append it to the existing sequence (1.7) to obtain a sequence for $k = 3$,

$$0 \to \ker(Q) \hookrightarrow \mathcal{D}^{1 \times m} \xrightarrow{Q} \mathcal{D}^{1 \times g} \xrightarrow{R} \mathcal{D}^{1 \times q} \to \mathcal{M} \to 0.$$

Iteratively, one proceeds with this computation of MLAs or, equivalently, syzygy modules, to obtain a sequence (1.5) of arbitrary length k. The module M_k in (1.5) is free for one such sequence iff it is free for all such sequences of the same length. In that case, (1.5) is called a **finite free resolution** of $\mathcal{M}$, and $\mathcal{M}$ is said to have projective dimension at most k. By the projective dimension of a behavior $\mathcal{B} = \ker(R(\sigma_1, \ldots, \sigma_r))$, we mean the projective dimension of the associated module $\mathcal{M} = \operatorname{cok}(R)$ as above. This notion is well-defined, as different kernel representations of $\mathcal{B}$ yield the same $\mathcal{M}$ [48]. Hilbert's syzygy theorem [39, p. 208] implies that for $\mathcal{D}$ as above, the projective dimension of a finitely generated $\mathcal{D}$-module cannot exceed r, the number of variables in $z = (z_1, \ldots, z_r)$. For $r = 1$, it is therefore clear that the projective dimension of $\mathcal{M}$ is at most one, and for $r = 2$, we have the desired result at least for those behaviors that are minimal in their transfer class [48]; see Section 2.6. This is not true for $r > 2$, however, as can be seen from the following example.

Example 1.2.2. Consider the behavior

$$\mathcal{B} = \{w : \mathbb{N}^3 \to \mathbb{R}^3, \exists l : \mathbb{N}^3 \to \mathbb{R} \text{ such that } w_i = \sigma_i l \text{ for } i = 1, 2, 3\}.$$

In detail, the defining equation for $w_1 : \mathbb{N}^3 \to \mathbb{R}$ reads

$$w_1(n_1, n_2, n_3) = l(n_1 + 1, n_2, n_3) \quad \text{for all } (n_1, n_2, n_3) \in \mathbb{N}^3.$$

The elimination of latent variables (here: l) will be discussed several times later in this work. In the present example, it yields

$$\mathcal{B} = \{w : \mathbb{N}^3 \to \mathbb{R}^3, \begin{bmatrix} 0 & -\sigma_3 & \sigma_2 \\ \sigma_3 & 0 & -\sigma_1 \\ -\sigma_2 & \sigma_1 & 0 \end{bmatrix} \begin{bmatrix} w_1 \\ w_2 \\ w_3 \end{bmatrix} = 0\}.$$

This is nothing but a discrete version of the fact that a vector field w is derivable from a scalar potential iff $\operatorname{curl}(w) = 0$. For the continuous time version, see Section 1.5. Although the rank of the kernel representation matrix is two (and this is an invariant of the behavior $\mathcal{B}$, *i.e.*, independent of the particular choice of the kernel representation matrix), the behavior $\mathcal{B}$ cannot be expressed as the solution set of less than three equations. This discrepancy is due to the fact that the projective dimension of $\mathcal{B}$ is two. Note that this phenomenon occurs although $\mathcal{B}$ is indeed minimal in its transfer class (which is equivalent to $\mathcal{B}$ having an image representation, or controllability in a sense to be defined below).

Due to these considerations, the concept of factor primeness has to be generalized as follows [48, p. 142].

Definition 3 *A matrix R is called* **factor left prime in the generalized sense (GFLP)** *if the existence of a factorization $R = DR_1$ (D not necessarily square) with* $\operatorname{rank}(R) = \operatorname{rank}(R_1)$ *implies the existence of a polynomial matrix E such that $R_1 = ER$.*

Note that this notion does not require full row rank. A test for GFLP is given by the following procedure.

Algorithm 1 *1. Solve the linear system of equations $R\zeta = 0$ over the polynomial ring, that is, find an integer m and a matrix $M \in \mathcal{D}^{q \times m}$ whose columns generate the syzygy module of R, i.e.,*

$$\{\zeta \in \mathcal{D}^q,\ R\zeta = 0\} = M\mathcal{D}^m = \operatorname{CM}(M).$$

In other words, construct a minimal right annihilator of R or, an exact sequence

$$\mathcal{D}^m \xrightarrow{M} \mathcal{D}^q \xrightarrow{R} \mathcal{D}^g.$$

2. Now find a minimal left annihilator of M, that is, a matrix $R^c \in \mathcal{D}^{g_c \times q}$ such that

$$\{\eta \in \mathcal{D}^{1 \times q},\ \eta M = 0\} = \operatorname{RM}(R^c).$$

We have two sequences

$$\mathcal{D}^{1 \times g} \xrightarrow{R} \mathcal{D}^{1 \times q} \xrightarrow{M} \mathcal{D}^{1 \times m}$$
$$\mathcal{D}^{1 \times g_c} \xrightarrow{R^c}$$

where the lower sequence is exact. The upper sequence is a complex ($RM = 0$), but not necessarily exact. Note that the mappings $\xi \mapsto M\xi$ and $\eta \mapsto \eta M$ are both denoted by M, but they can always be distinguished by their domain and co-domain.

3. Fix a term order and compute the row Gröbner matrices of R and R^c, $\operatorname{rG}(R)$ and $\operatorname{rG}(R^c)$ to check whether the rows of the two matrices generate the same module. If yes, also the upper sequence in Step 2 above is exact, i.e., R is a minimal left annihilator of M.

In Step 1 above, it suffices to solve $R\zeta = 0$ over the field $\mathcal{K}$, that is, the problem can be reduced to linear algebra: We merely compute a basis of the solution space of a system of linear equations over a field, *i.e.*,

$$\{\zeta \in \mathcal{K}^q, \ R\zeta = 0\} = M\mathcal{K}^{q-\mathrm{rank}(R)}.$$

Without loss of generality, we may assume that the entries of M are polynomials instead of rational functions; otherwise, each column can be multiplied by a common multiple of all denominator polynomials appearing in it. Step 2 however depends crucially on the computation of MLAs, or syzygy modules over $\mathcal{D}$. This is based on a computer algebraic technique, as mentioned above.

Lemma 1.2.3. *The matrix R is GFLP iff the procedure described above yields equality of the row Gröbner matrices* $\mathrm{rG}(R)$ *and* $\mathrm{rG}(R^c)$.

Proof. From the algorithm, we have $RM = 0$ and $\mathrm{rank}\,(R) = \mathrm{rank}\,(R^c)$. Since R^c is a minimal left annihilator of M, there exists a polynomial matrix D such that $R = DR^c$. If R is GFLP, we also have $R^c = ER$, hence $\mathrm{RM}(R) = \mathrm{RM}(R^c)$, which is equivalent to the equality of the row Gröbner matrices.

Conversely, let $R = DR_1$ be any factorization with $\mathrm{rank}\,(R) = \mathrm{rank}\,(R_1)$. Then $R\zeta = 0$ iff $R_1\zeta = 0$ for $\zeta \in \mathcal{K}^q \supset \mathcal{D}^q$. Thus $R_1 M = 0$. But R^c is a minimal left annihilator of M, from which we conclude that $R_1 = YR^c$ for some polynomial matrix Y. On the other hand, from $\mathrm{rG}(R) = \mathrm{rG}(R^c)$ we have $R^c = ZR$ for some polynomial matrix Z and hence $R_1 = YZR$. $\qquad\square$

1.2.3 Relations Between the Various Primeness Notions

It is immediate that ZLP implies GFLP as $R = DR_1$ with $\mathrm{rank}\,(R) = \mathrm{rank}\,(R_1)$ always implies the existence of a *rational* matrix E such that $R_1 = ER$. Now if R possesses a polynomial right inverse X, we conclude that $E = R_1 X$, hence E is polynomial.

Also note that GFLP implies FLP in the case of a full row rank matrix as it postulates the existence of a polynomial inverse E of D, thus unimodularity of D. Conversely, FLP in the classical sense does not necessarily imply GFLP, as can be seen from the following example.

Example 1.2.3. Consider again the matrix

$$R = \begin{bmatrix} z_1 & 0 & -z_2 \\ 0 & z_1 & z_3 \end{bmatrix} \tag{1.8}$$

whose factor left primeness was shown in an example above. Computing a minimal right annihilator yields

$$M = \begin{bmatrix} z_2 \\ -z_3 \\ z_1 \end{bmatrix}$$

and an MLA of M is

$$R^c = \begin{bmatrix} z_1 & 0 & -z_2 \\ 0 & z_1 & z_3 \\ z_3 & z_2 & 0 \end{bmatrix}$$

whose row module is strictly greater than that of R itself. This shows that R is not GFLP.

Wood *et al.* [83] proved that, despite appearances, GFLP is a generalization of minor primeness rather than of factor primeness, in the sense that MLP and GFLP coincide in the full row rank case. If the notion of minor primeness is generalized to representations with rank deficiencies, it is at least necessary for GFLP; see Theorem 2 below.

Let $R \in \mathcal{D}^{g \times q}$. Let ρ denote the rank of R and consider the $\rho \times \rho$ minors of R. We define modified versions of zero, weak zero, and minor primeness by adapting Definition 1 to this situation (*i.e.*, we consider the minors of order ρ instead of g). Note that the distinction between right and left primeness does not make sense in this new setting. Also, we no longer need to require that $g \leq q$. Let $\mathcal{I}(R)$ be the ideal generated by the minors of R of order equal to the rank of R. Define $\mathcal{I}(0) := \langle 1 \rangle = \mathcal{D}$. The characterization of the three primeness notions in terms of the dimension of $\mathcal{I}(R)$, given in Corollary 1, is still valid. In particular, R is minor prime iff $\mathrm{codim}(\mathcal{I}(R)) \geq 2$, where we define $\mathrm{codim}(\mathcal{I}) = r - \dim(\mathcal{I})$ for proper ideals

$$\mathcal{I} \subset \mathcal{D} = \mathbb{R}[z_1, \ldots, z_r],$$

and by convention, $\mathrm{codim}(\mathcal{D}) = \infty$. Thus the co-dimension measures the "size" of a polynomial ideal in the sense that

$$\mathcal{I}_1 \subseteq \mathcal{I}_2 \quad \Rightarrow \quad \mathrm{codim}(\mathcal{I}_1) \leq \mathrm{codim}(\mathcal{I}_2).$$

The following criterion for exactness of a complex of finitely generated free modules over $\mathcal{D}$ is a special case of a theorem of Buchsbaum and Eisenbud [13] (see also [18, p. 500], [47, p. 193]).

Theorem 1 *[13] Let*

$$0 \longrightarrow M_n \xrightarrow{F_n} M_{n-1} \longrightarrow \ldots \longrightarrow M_1 \xrightarrow{F_1} M_0 \qquad (1.9)$$

be a complex of finitely generated free $\mathcal{D}$-modules, i.e., $F_i F_{i+1} = 0$ for $1 \leq i \leq n$ (put $F_{n+1} := 0$). The complex is exact iff, for $1 \leq i \leq n$,

1. $\mathrm{rank}\,(F_{i+1}) + \mathrm{rank}\,(F_i) = \mathrm{rank}\,(M_i)$;
2. $\mathrm{codim}(\mathcal{I}(F_i)) \geq i$.

Condition 2 is formulated in terms of the depth (grade) of ideals in [18], [47], but for polynomial ideals, this quantity coincides with the co-dimension (this is true over any Cohen–Macaulay ring).

Theorem 2 *Let $R \in \mathcal{D}^{g \times q}$. Then R is generalized factor (left/right) prime only if it is minor prime, that is, only if $\mathrm{codim}(\mathcal{I}(R)) \geq 2$, where $\mathcal{I}(R)$ denotes the ideal generated by the minors of R of order equal to the rank of R.*

Proof. Without loss of generality, assume that R is GFLP (if not, replace R by its transpose). Then there exists an exact sequence

$$\mathcal{D}^{1 \times g} \xrightarrow{R} \mathcal{D}^{1 \times q} \xrightarrow{M} \mathcal{D}^{1 \times m}. \tag{1.10}$$

Let

$$0 \to \dots \to \mathcal{D}^{1 \times g} \xrightarrow{R} \mathcal{D}^{1 \times q} \to \mathcal{M} = \mathcal{D}^{1 \times q} / \mathcal{D}^{1 \times g} R \to 0$$

be a finite free resolution of $\mathcal{M}$, which exists due to Hilbert's syzygy theorem [39, p. 208]. Combining this with the exact sequence (1.10), we obtain an exact sequence of finitely generated free $\mathcal{D}$-modules

$$0 \to \dots \to \mathcal{D}^{1 \times g} \xrightarrow{R} \mathcal{D}^{1 \times q} \xrightarrow{M} \mathcal{D}^{1 \times m}.$$

Now Theorem 1 implies that $\mathrm{codim}(\mathcal{I}(R)) \geq 2$, hence R is minor prime. $\square$

The converse of Theorem 2 does not hold, *i.e.*, minor primeness is not sufficient for GFLP. This fact is illustrated by the following example.

Example 1.2.4. The matrix

$$R = \begin{bmatrix} z_1 & z_1 \\ z_2 & z_2 \end{bmatrix}$$

is minor prime, as $\mathrm{rank}\,(R) = 1$ and $\mathcal{I}(R) = \langle z_1, z_2 \rangle \subset \mathbb{R}[z_1, z_2]$ has codimension two. But an MRA of R is

$$M = \begin{bmatrix} 1 \\ -1 \end{bmatrix}$$

and an MLA of M is $R^c = \begin{bmatrix} 1 & 1 \end{bmatrix}$, hence R is not GFLP.

Summing up, we have the following relations in the full row rank case:

$$\mathrm{ZLP} \Rightarrow \mathrm{WLP} \Rightarrow \dots \Rightarrow \mathrm{MLP} \tag{1.11}$$

and

$$\mathrm{MLP} \Leftrightarrow \mathrm{GFLP} \Rightarrow \mathrm{FLP},$$

(analogously for right primeness), whereas in the general case we only have

$$\text{zero prime} \Rightarrow \text{weakly zero prime} \Rightarrow \dots \Rightarrow \text{minor prime} \tag{1.12}$$

and

$$\text{zero prime} \Rightarrow \text{generalized factor (left/right) prime} \Rightarrow \text{minor prime.} \quad (1.13)$$

The dots in (1.11) and (1.12) refer to the gap in the corresponding ideal dimensions $-1, 0, \ldots, r-2$, or co-dimensions $\infty, r, \ldots, 2$, respectively. Wood *et al.* [83] described the "missing links" in (1.11), that is, the various primeness degrees in between. The first implication in (1.13) remains to be shown; this will be done in the subsequent section.

1.2.4 Module–theoretic Interpretation of Primeness

A polynomial matrix $R \in \mathcal{D}^{g \times q}$ can be interpreted as the kernel representation of a linear shift-invariant behavior

$$\mathcal{B} = \ker(R) = \ker(R(\sigma_1, \ldots, \sigma_r)) = \{w : \mathbb{N}^r \to \mathbb{R}^q, \ R(\sigma_1, \ldots, \sigma_r)w = 0\},$$

the action of R on w being given by the shifts σ_i. There are many different kernel representations of one and the same behavior, but according to Oberst's duality [48], they all generate the same row module. Thus the question arises naturally whether the primeness notions discussed above have module–theoretic interpretations in terms of $\mathcal{M} = \mathcal{D}^{1 \times q}/\mathcal{D}^{1 \times g}R = \mathcal{D}^{1 \times q}/\mathrm{RM}(R) = \mathrm{cok}(R)$, where R is interpreted as a mapping $\mathcal{D}^{1 \times g} \to \mathcal{D}^{1 \times q}$, $\eta \mapsto \eta R$ (multiplication from the left).

We know that zero primeness of R corresponds to $\mathcal{I}(R) = \langle 1 \rangle = \mathcal{D}$. A well-known lemma [13] implies that this is equivalent to the module $\mathcal{M}$ being projective. By the Theorem of Quillen and Suslin (also known as Serre's conjecture [39, p. 116/117]) projective modules over the polynomial ring $\mathcal{D}$ are already free. We have the following correspondence between a representation matrix R and $\mathcal{M} = \mathrm{cok}(R)$:

$$R \text{ zero prime} \Leftrightarrow \mathcal{M} \text{ free.}$$

A full column rank matrix R is ZRP iff it has a polynomial left inverse (Bézout relation $YR = I_q$), hence $\mathrm{RM}(R) = \mathcal{D}^{1 \times g}R = \mathcal{D}^{1 \times q}$, or $\mathcal{M} = 0$. As a mapping, R is surjective. Thus

$$R \text{ ZRP (zero prime and full column rank)} \Leftrightarrow \mathcal{M} = 0.$$

On the other hand, GFLP has been characterized by Oberst [48, p. 142] as follows:

$$R \text{ generalized factor left prime} \Leftrightarrow \mathcal{M} \text{ torsion-free.} \quad (1.14)$$

As a free module is torsion-free, we obtain the implication

$$\text{zero prime} \Rightarrow \text{GFLP}$$

of (1.13) and the corresponding assertion for GFRP is obtained by transposition.

1.3 Controllability, Observability, and Autonomy

Let $\mathcal{A} = \mathbb{R}^{\mathbb{N}^r}$. The elements of $\mathcal{A}$ are r-fold indexed real sequences. As **signals**, we will consider vectors of such sequences, *i.e.*, the signals will be in $\mathcal{A}^q$, where $q \geq 1$ is an integer. In terms of Willems' original definition of a behavioral system, $T = \mathbb{N}^r$ is our "time domain," or rather "signal domain," as the interpretation as time usually makes sense only in the 1D setting. The signal value space is $\mathbb{R}^q$ and our signals $w \in \mathcal{A}^q$ are functions

$$ w : \mathbb{N}^r \rightarrow \mathbb{R}^q. $$

Finally, the **behavior** $\mathcal{B} \subseteq \mathcal{A}^q$ of the system is determined by certain system laws. In the linear and "time-invariant" (the term "shift-invariant" is more adequate in the multidimensional case), these will be constant-coefficient linear partial difference equations. Thus, let

$$ \mathcal{B} = \ker(R) = \{ w \in \mathcal{A}^q, \; R(\sigma)w := R(\sigma_1, \ldots, \sigma_r)w = 0 \} $$

be a behavior in **kernel representation** with a polynomial matrix $R \in \mathcal{D}^{g \times q}$ (not necessarily with $g \leq q$) in which the indeterminates z_i are replaced by the shift operators σ_i defined by

$$ (\sigma_i w)(t_1, \ldots, t_r) = w(t_1, \ldots, t_{i-1}, t_i + 1, t_{i+1}, \ldots, t_r). $$

For any $m \in \mathbb{N}^r$, we write σ^m for the compound shift $\sigma_1^{m_1} \cdots \sigma_r^{m_r}$, and thus

$$ (\sigma^m w)(n) = (\sigma_1^{m_1} \cdots \sigma_r^{m_r} w)(n_1, \ldots, n_r) = w(n + m) $$

for $n, m \in \mathbb{N}^r$ and $w \in \mathcal{A}^q$.

We say that R is a minimal left annihilator (MLA) of $M \in \mathcal{D}^{q \times m}$ iff

$$ \{ \eta \in \mathcal{D}^{1 \times q}, \; \eta M = 0 \} = \mathrm{RM}(R). \tag{1.15} $$

The notion has already appeared in the algorithm for testing a matrix for the GFLP property. The so-called **fundamental principle** [48, p. 23] will be of central importance in the following: A linear system of inhomogeneous partial difference equations $M(\sigma)l = w$, where the right side is assumed to be given, is solvable iff $R(\sigma)w = 0$ for a minimal left annihilator R of M. In other words, (1.15) implies

$$ \exists l : M(\sigma)l = w \quad \Leftrightarrow \quad R(\sigma)w = 0. \tag{1.16} $$

Thus, if R is a MLA of M, we have

$$ \mathcal{B} = \ker(R) = \mathrm{im}(M) = \{ w \in \mathcal{A}^q, \; \exists \, l \in \mathcal{A}^m \text{ such that } w = M(\sigma)l \}, $$

i.e., there exists an **image representation** of $\mathcal{B}$. Oberst's duality theorem [48] implies that (1.15) and (1.16) are in fact equivalent for the signal spaces considered here. Moreover, we have $\mathcal{B}_1 = \ker(R_1) \subseteq \mathcal{B}_2 = \ker(R_2)$ iff $\mathrm{RM}(R_1) \supseteq \mathrm{RM}(R_2)$, that is, iff $R_2 = X R_1$ for some polynomial matrix X [48, p. 35].

Theorem 3 *If R is GFLP, then there exists an image representation*

$$B = \text{im}(M) = \{w \in \mathcal{A}^q, \exists\, l \in \mathcal{A}^m \text{ with } w = M(\sigma)l\},$$

where the polynomial matrix M is constructed by Algorithm 1. Conversely, if B possesses an image representation, any kernel representation matrix of B is GFLP.

Proof. From Algorithm 1 and the lemma following it, we have $\text{RM}(R) = \text{RM}(R^c)$ and hence $B = \ker(R) = \ker(R^c)$ for a GFLP matrix R. From the construction of R^c and M, and from the fundamental principle, it follows that $\ker(R^c) = \text{im}(M)$.

Conversely, if $B = \text{im}(M') = \ker(R)$, then R is a minimal left annihilator of M', in particular, $RM' = 0$ and $M' = MX$ for some polynomial matrix X, as M is a MRA of R. In order to show that R is GFLP, we need to prove that $B = \ker(R^c)$, that is, R^c should be a minimal left annihilator of M'. But $R^c M' = R^c M X = 0$ and if $Y M' = 0$, then $Y = ZR$. On the other hand, we always have $R = DR^c$ for some polynomial matrix D, hence $Y = ZDR^c$ follows.   □

Definition 4 *The preceding theorem suggests calling a behavior* **controllable** *if it admits a kernel representation with a GFLP matrix.*

A justification of this purely algebraic definition of controllability will be given in Section 1.4, where it will be characterized in terms of concatenability of system trajectories, as should be expected from a behavioral controllability notion.

Lemma 1.3.1. *The matrix R^c from the algorithm defines the controllable part of $B = \ker(R)$, that is, $B^c := \ker(R^c) \subseteq \ker(R)$ is the largest controllable subsystem of B.*

Proof. By the construction of R^c and M, we clearly have $\text{RM}(R) \subseteq \text{RM}(R^c)$, hence $B^c \subseteq B$, and $\ker(R^c) = \text{im}(M)$, hence controllability of B^c. It remains to be shown that B^c is the largest subsystem of this type. Assume that $B_1 \subseteq B$ is controllable. Then $B_1 = \text{im}(M_1) = \ker(R_1) \subseteq \ker(R)$, particularly, $RM_1 = 0$, hence $M_1 = MX$ for some polynomial matrix X. If $\eta \in \text{RM}(R^c)$, then $\eta M = 0$ and $\eta M_1 = 0$, thus $\eta \in \text{RM}(R_1)$. This proves $B_1 \subseteq B^c$.   □

Definition 5 *A* **cw-ideal** *is a non-empty subset $I \subseteq \mathbb{N}^r$ that satisfies*

$$n \in I \;\Rightarrow\; n + m \in I \quad \text{for all } m \in \mathbb{N}^r.$$

A behavior B is called **autonomous** *if there exists a subset $J \subset \mathbb{N}^r$ such that $\mathbb{N}^r \setminus J$ is a cw-ideal, and any trajectory $w \in B$ is uniquely determined by $w|_J = (w(j))_{j \in J}$, that is,*

$$w|_J = 0 \;\Rightarrow\; w = 0.$$

The dimension of $\mathcal{B}$ as a real vector space cannot exceed the cardinality of such a set J. The behavior $\mathcal{B}$ is **finite-dimensional** *(as a real vector space) if there exists a finite set J with these properties. For $1 \leq i \leq q$, let $\pi_i : \mathcal{B} \to \mathcal{A}$ denote the projection of $\mathcal{B} \subset \mathcal{A}^q$ onto the i-th component. The i-th variable of $\mathcal{B}$ is called* **free** *if π_i is surjective.*

Lemma 1.3.2. *The following assertions are equivalent:*

1. *$\mathcal{B}$ is autonomous;*
2. *$\mathcal{B}$ has no free variables;*
3. *Any kernel representation matrix of $\mathcal{B}$ has full column rank.*

Proof. It is easy to see that the existence of a free variable implies that $\mathcal{B}$ is not autonomous. If $\mathcal{B} = \ker(R)$ and R does not have full column rank, then there exists a column of R that can be written as a rational combination of the remaining columns, say $R_{-q} = [R_{-1} \ldots R_{-(q-1)}]X$ for some rational matrix X. Then the variable w_q is free, since $R(\sigma)w = 0$ is solvable whatever $w_q \in \mathcal{A}$ due to the fact that any minimal left annihilator of the matrix consisting of the first $q-1$ columns also annihilates the last column. If $\mathcal{B} = \ker(R)$ with a full column rank matrix R, then according to Oberst [48, p. 93] a decomposition

$$\mathcal{A} = \pi_i \mathcal{B} \oplus \mathbb{R}^{I(i)}$$

with a cw-ideal $I(i)$ can be constructed for $1 \leq i \leq q$. Then $I := \cap_i I(i)$ is again a cw-ideal, and with $J := \mathbb{N}^r \setminus I$ it follows that $\mathcal{B}$ is autonomous.   □

Lemma 1.3.3. *Let $\mathcal{B} = \ker(R)$ be autonomous. Then $\mathcal{B} = \{0\}$ iff R is zero right prime and $\mathcal{B}$ is finite-dimensional iff R is weakly zero right prime.*

Proof. $\mathcal{B} = \{0\}$ iff $\mathrm{RM}(R) = \mathcal{D}^{1 \times q}$, that is, iff the Bézout equation $YR = I_q$ is solvable. $\mathcal{B}$ is finite-dimensional iff in the decomposition

$$\mathcal{A}^q = \mathcal{B} \oplus \prod_i \mathbb{R}^{I(i)}$$

the sets $I(i)$ are complements of finite sets. By duality, this is equivalent to

$$\mathcal{D}^{1 \times q}/\mathrm{RM}(R)$$

being a finite-dimensional real vector space.   □

Definition 6 *Let I be a subset of $q = \{1, \ldots, q\}$ and for a trajectory $w \in \mathcal{B}$, let $w_1 := (w_i)_{i \in I}$ and $w_2 := (w_i)_{i \notin I}$. The sub-trajectory w_1 is called* **strongly observable** *from w_2 if $w_2 = 0$ implies $w_1 = 0$ and* **weakly observable** *from w_2 if there exists a finite-dimensional behavior $\mathcal{B}_1$ such that $w_2 = 0$ implies $w_1 \in \mathcal{B}_1$.*

Corollary 2 *Rewrite $R(\sigma)w = 0$ as $R_1(\sigma)w_1 = R_2(\sigma)w_2$. Then w_1 is strongly observable from w_2 iff the polynomial matrix R_1 is zero right prime, and it is weakly observable from w_2 iff R_1 is weakly zero right prime.*

1.3.1 Controllable–autonomous Decompositions

It is well-known that a 1D behavior $\mathcal{B}$ admits a direct sum decomposition [32]

$$\mathcal{B} = \mathcal{B}^c \oplus \mathcal{B}^a$$

where $\mathcal{B}^c$ denotes the controllable part of $\mathcal{B}$ ("forced motion") and $\mathcal{B}^a$ is autonomous ("free motion"). The following theorem treats the general r-dimensional situation. It turns out that controllable–autonomous decompositions still exist for multidimensional systems (as they correspond to the torsion/torsion-free decomposition of the corresponding modules [84]), but they are, in general, not direct. For 2D systems, this has already been pointed out in [23]. Necessary and sufficient conditions for the existence of direct sum decompositions are given in Theorem 5.

Theorem 4 *There exist subsystems $\mathcal{B}^c$ and $\mathcal{B}^a$ of $\mathcal{B}$ such that*

1. *$\mathcal{B}^c$ is controllable,*
2. *$\mathcal{B}^a$ is autonomous, and*
3. *$\mathcal{B} = \mathcal{B}^c + \mathcal{B}^a$.*

Proof. Let $\mathcal{B}^c = \ker(R^c)$ be the controllable part of $\mathcal{B}$ with $R = DR^c$. Let Π be a permutation matrix such that

$$R^c \Pi = \begin{bmatrix} R_1^c & R_2^c \end{bmatrix}$$

with a matrix R_1^c of full column rank and $\mathrm{rank}\,(R_1^c) = \mathrm{rank}\,(R^c)$. Then $R_2^c = R_1^c X$ for some rational matrix X. Without loss of generality, assume that

$$R^c = \begin{bmatrix} R \\ R' \end{bmatrix}$$

with $R' = YR$ for some rational matrix Y. Hence with $R\Pi = DR^c\Pi = D[R_1^c R_2^c]$, the matrix $R_1 := DR_1^c$ has full column rank. Define

$$\mathcal{B}^a := \left\{ w \in \mathcal{A}^q, \begin{bmatrix} R_1(\sigma) & 0 \\ 0 & I \end{bmatrix} \Pi^{-1} w = 0 \right\}$$

which is autonomous and contained in $\mathcal{B}$. It remains to be proven that $\mathcal{B}$ is contained in $\mathcal{B}^c + \mathcal{B}^a$. Let $w \in \mathcal{B}$ be given. Rewrite it as

$$w = \left(w - \Pi \begin{bmatrix} v \\ 0 \end{bmatrix} \right) + \Pi \begin{bmatrix} v \\ 0 \end{bmatrix},$$

where v is a solution of $R_1^c(\sigma)v = R^c(\sigma)w$. This system of equations is solvable, since a minimal left annihilator of R_1^c also annihilates R^c. By the construction of v, the first summand of w is contained in $\mathcal{B}^c$, and the second summand is contained in $\mathcal{B}^a$, as $R_1(\sigma)v = D(\sigma)R_1^c(\sigma)v = D(\sigma)R^c(\sigma)w = R(\sigma)w = 0$. $\qquad\square$

Unlike the 1D case, $\mathcal{B}^c \cap \mathcal{B}^a \neq \{0\}$ in general. It has been conjectured that

$$\mathcal{B} = \mathcal{B}^c \oplus \mathcal{B}^a \qquad (1.17)$$

iff $\mathcal{B}^c$ is even strongly controllable, i.e., iff R^c is zero left prime. However, it has been pointed out by Valcher [75] that even for 2D systems, the zero left primeness of R^c is sufficient, but not necessary for the existence of a direct sum decomposition (1.17). In order to derive necessary and sufficient conditions for (1.17), the following lemma is useful.

Lemma 1.3.4. *Let R_1, R_2 be zero right co-prime, that is, the block matrix $[\begin{array}{cc} R_1^T & R_2^T \end{array}]^T$ is zero right prime. Then there exist polynomial matrices X and Y such that*

$$\begin{bmatrix} R_1 & -X \\ R_2 & Y \end{bmatrix}$$

is unimodular. Let the polynomial matrices T, U, V, W be defined by

$$\begin{bmatrix} R_1 & -X \\ R_2 & Y \end{bmatrix}^{-1} = \begin{bmatrix} T & U \\ -V & W \end{bmatrix} \qquad (1.18)$$

with the obvious partition (i.e., $R_1 T$ a square matrix). Then $[\begin{array}{cc} -V, & W \end{array}]$ is a MLA of $\begin{bmatrix} R_1 \\ R_2 \end{bmatrix}$, and $[\begin{array}{ccc} V, & W, & -V R_1 \end{array}]$ is a MLA of

$$M = \begin{bmatrix} R_1 & 0 \\ 0 & R_2 \\ I & I \end{bmatrix}.$$

Proof. The first statement follows from the Quillen–Suslin Theorem [89]. Let $[\begin{array}{cc} -A, & B \end{array}]$ be any left annihilator of $\begin{bmatrix} R_1 \\ R_2 \end{bmatrix}$, then

$$\{\begin{array}{cc} -A & B \end{array}\} \begin{bmatrix} R_1 & -X \\ R_2 & Y \end{bmatrix} = [\begin{array}{cc} 0 & Z \end{array}]$$

where $Z = AX + BY$. Then

$$[\begin{array}{cc} -A & B \end{array}] = [\begin{array}{cc} 0 & Z \end{array}] \begin{bmatrix} T & U \\ -V & W \end{bmatrix} = Z [\begin{array}{cc} -V & W \end{array}].$$

Finally, let $[\begin{array}{ccc} A, & B, & -C \end{array}]$ be a left annihilator of M, then $C = AR_1 = BR_2$, hence $A = ZV$, $B = ZW$, and $C = ZVR_1$. □

Now the following theorem characterizes the sub-behaviors $\mathcal{B}_1$ of $\mathcal{B}$ that are direct summands of $\mathcal{B}$. Note that no assumption is made on the primeness of R_1.

Theorem 5 *Let $\mathcal{B}_1$ and $\mathcal{B}$ be behaviors, and let $R_1 \in \mathcal{D}^{g_1 \times q}$ be a kernel representation matrix of $\mathcal{B}_1$.*

The following are equivalent:

1. $\mathcal{B}_1$ *is a direct summand of $\mathcal{B}$, that is, $\mathcal{B} = \mathcal{B}_1 \oplus \mathcal{B}_2$ for some behavior $\mathcal{B}_2$;*
2. *There exists, for some integer $g \geq 1$, a polynomial matrix $V \in \mathcal{D}^{g \times g_1}$ such that*
   a) $\mathcal{B} = \ker(V R_1)$
   b) *V and R_1 are zero skew prime, that is, there exist polynomial matrices $T \in \mathcal{D}^{q \times g_1}$ and $X \in \mathcal{D}^{g_1 \times g}$ such that*

$$R_1 T + XV = I. \tag{1.19}$$

3. *There exists, for some integer $g \geq 1$, a polynomial matrix $V \in \mathcal{D}^{g \times g_1}$ such that*
   a) $\mathcal{B} = \ker(V R_1)$
   b) *for some $W \in \mathcal{D}^{g \times g_2}$, $R_2 \in \mathcal{D}^{g_2 \times q}$, where $g_2 = q + g - g_1$, we have*

$$V R_1 = W R_2$$

   with V, W zero left co-prime and R_1, R_2 zero right co-prime.

Proof. "$1 \Rightarrow 2$": Let R_2 be a kernel representation matrix of $\mathcal{B}_2$. By assumption,

$$\begin{bmatrix} R_1 \\ R_2 \end{bmatrix}$$

is zero right prime (*i.e.*, $\mathcal{B}_1 \cap \mathcal{B}_2 = \{0\}$), and $w \in \mathcal{B}$ iff $w = l_1 + l_2$ with $l_1 \in \mathcal{B}_1$, $l_2 \in \mathcal{B}_2$. Hence

$$w \in \mathcal{B} \quad \Leftrightarrow \quad \exists l_1, l_2 : \quad \begin{bmatrix} 0 \\ 0 \\ w \end{bmatrix} = \begin{bmatrix} R_1 & 0 \\ 0 & R_2 \\ I & I \end{bmatrix} \begin{bmatrix} l_1 \\ l_2 \end{bmatrix} =: M \begin{bmatrix} l_1 \\ l_2 \end{bmatrix}. \tag{1.20}$$

Let $\begin{bmatrix} V, & W, & -V R_1 \end{bmatrix}$ be the MLA of M constructed in Lemma 1.3.4, then it follows by the fundamental principle [48],

$$w \in \mathcal{B} \quad \Leftrightarrow \quad \begin{bmatrix} V & W & -V R_1 \end{bmatrix} \begin{bmatrix} 0 \\ 0 \\ w \end{bmatrix} = 0 \quad \Leftrightarrow \quad V R_1 w = 0.$$

Hence $\mathcal{B} = \ker(V R_1)$. From (1.18), we have

$$R_1 T + XV = I.$$

"$2 \Rightarrow 3$": By assumption, T and V are zero right co-prime. Let $\tilde{U}$ and $\tilde{W}$ be such that

$$\begin{bmatrix} T & \tilde{U} \\ -V & \tilde{W} \end{bmatrix}$$

is unimodular. Define the $(q + g - g_1) \times q$ polynomial matrix R_2 by

$$\begin{bmatrix} P & Q \\ R_2 & Y \end{bmatrix} = \begin{bmatrix} T & \tilde{U} \\ -V & \tilde{W} \end{bmatrix}^{-1}$$

and consider

$$\begin{bmatrix} R_1 & -X \\ R_2 & Y \end{bmatrix} \begin{bmatrix} T & \tilde{U} \\ -V & \tilde{W} \end{bmatrix} = \begin{bmatrix} I & Z \\ 0 & I \end{bmatrix}$$

where $Z = R_1 \tilde{U} - X \tilde{W}$. This implies that the left-most matrix is unimodular, in particular, R_1 and R_2 are zero right co-prime, and

$$\begin{bmatrix} R_1 & -X \\ R_2 & Y \end{bmatrix}^{-1} = \begin{bmatrix} T & \tilde{U} \\ -V & \tilde{W} \end{bmatrix} \begin{bmatrix} I & -Z \\ 0 & I \end{bmatrix} = \begin{bmatrix} T & U \\ -V & W \end{bmatrix}$$

where $U = \tilde{U} - TZ$ and $W = VZ + \tilde{W}$. Now $VR_1 = WR_2$ and V, W are zero left co-prime.

"3 $\Rightarrow$ 1": Let $\tilde{X}$, $\tilde{Y}$ be such that

$$V\tilde{X} + W\tilde{Y} = I$$

and let T, U be such that

$$TR_1 + UR_2 = I.$$

Then

$$\begin{bmatrix} T & U \\ -V & W \end{bmatrix} \begin{bmatrix} R_1 & -\tilde{X} \\ R_2 & \tilde{Y} \end{bmatrix} = \begin{bmatrix} I & Z \\ 0 & I \end{bmatrix}$$

where $Z = -T\tilde{X} + U\tilde{Y}$. Due to the assumption on g_2, the matrices on the left are square. Thus

$$\begin{bmatrix} T & U \\ -V & W \end{bmatrix}^{-1} = \begin{bmatrix} R_1 & -\tilde{X} \\ R_2 & \tilde{Y} \end{bmatrix} \begin{bmatrix} I & -Z \\ 0 & I \end{bmatrix} = \begin{bmatrix} R_1 & -X \\ R_2 & Y \end{bmatrix}$$

where $X = R_1 Z + \tilde{X}$ and $Y = \tilde{Y} - R_2 Z$. According to Lemma 1.3.4, $[\, V, \ W, \ -VR_1 \,]$ is a MLA of M, but as $\mathcal{B} = \ker(VR_1)$, this implies (1.20) and hence $\mathcal{B} = \mathcal{B}_1 \oplus \mathcal{B}_2$, where $\mathcal{B}_2 := \ker(R_2)$. $\square$

Note that the existence of V with $\mathcal{B} = \ker(VR_1)$ merely amounts to the requirement $\mathcal{B}_1 \subseteq \mathcal{B}$, but V is *essentially* non-unique for systems of dimension greater than two (*i.e.*, not "unique up to unimodular right factors"). Theorem 5, when applied to the case $\mathcal{B}_1 = \mathcal{B}^c$, shows clearly that zero left primeness of R^c is sufficient, but not necessary for $\mathcal{B} = \mathcal{B}^c \oplus \mathcal{B}^a$ (if R^c is ZLP, condition (1.19) can be satisfied with $X = 0$).

1.4 Controllability as Concatenability of Trajectories[1]

Controllability is a concept fundamental to any system-theoretic paradigm. The definition usually given for 1D systems seems to be closely related to state-space representations, to the concepts of "input" and "state." The behavioral approach of Willems has introduced a new intuitive idea of controllability in terms of the system trajectories [81].

The definition of behavioral controllability given by Rocha in [59] is an extension of the original 1D definition to discrete systems defined on $\mathbb{Z}^2$, and generalizes directly to systems defined on $\mathbb{Z}^r$. For such systems, it is a highly natural definition, and leads to a variety of useful characterizations of controllability [59, 84]. The definition given in [59, 84] also makes sense in the context of discrete systems defined on $\mathbb{N}^r$, but in this case, the characterizations fail. Also, we have undesirable phenomena such as the existence of non-trivial behaviors which are both autonomous and controllable. This suggests that the definition is not suitable for systems on $\mathbb{N}^r$. Furthermore, Rocha's definition causes these problems even when applied to 1D systems on $\mathbb{N}$, which indicates that great care needs to be taken in the formulation of the 1D definition. As discussed for example by Rosenthal, Schumacher and York [64], a change in the signal domain can affect many important system-theoretic properties.

In this section we propose a new definition of behavioral controllability, which works equally well in the cases of systems defined on $\mathbb{Z}^r$ and on $\mathbb{N}^r$. The unified definition can also be applied to systems defined on "mixed" signal domains $\mathbb{Z}^{r_1} \times \mathbb{N}^{r_2}$. The new definition is equivalent to Rocha's definition in the $\mathbb{Z}^r$-case, but it admits the characterizations given in [84] for both classes of systems. In particular, this work establishes that for any discrete rD behavior, controllability is equivalent to minimality (in the transfer class).

1.4.1 Previous Definitions of Controllability

In this section, we will consider a slightly more general notion of a discrete rD behavior than previously: first of all, we will admit arbitrary signal value fields $\mathbb{F}$ (up to now: $\mathbb{R}$). Note that the presented results on multivariate polynomial matrix primeness, in particular, characterization (1.14), the fundamental principle (1.16), and Theorem 3, hold for arbitrary coefficient fields. Moreover, we have considered $T = \mathbb{N}^r$ up to now, and we have mentioned the case $T = \mathbb{Z}^r$ in several places. Mathematically, we deal with the ordinary polynomial ring, that is, $\mathbb{F}[z_1, \ldots, z_r]$, in the former, and with the ring of Laurent polynomials in the latter case, i.e., $\mathbb{F}[z_1, \ldots, z_r, z_1^{-1}, \ldots, z_r^{-1}]$. Observe that the set of units of the ordinary polynomial ring is $\mathbb{F} \setminus \{0\}$, whereas the units of the Laurent polynomial ring are all elements of the form

---

[1] The contents of this section are joint work with J. Wood [85].

$fz_1^{a_1} \cdots z_r^{a_r}$, where $a_i \in \mathbb{Z}$, and $0 \neq f \in \mathbb{F}$. Primeness results can easily be translated to the Laurent polynomial case by observing that [91]

$$\mathbb{F}[z_1, \dots, z_r, z_1^{-1}, \dots, z_r^{-1}] \cong \mathbb{F}[z_1, \dots, z_r, \zeta_1, \dots, \zeta_r]/\langle z_1\zeta_1 - 1, \dots, z_r\zeta_r - 1 \rangle.$$

Alternatively, one can argue that the Laurent polynomial ring is the **ring of fractions** of the polynomial ring with the multiplicatively closed denominator set

$$\{z_1^{a_1} \cdots z_r^{a_r}, \ (a_1, \dots, a_r) \in \mathbb{N}^r\}.$$

It is again a Noetherian unique–factorization–domain [39, p. 81]; moreover, with the natural inclusion

$$\mathbb{F}[z] := \mathbb{F}[z_1, \dots, z_r] \subset \mathbb{F}[z_1, \dots, z_r, z_1^{-1}, \dots, z_r^{-1}] =: \mathbb{F}[z, z^{-1}],$$

there is an injection of the set of Laurent polynomial ideals into the set of polynomial ideals, given by

$$\mathcal{I} \subseteq \mathbb{F}[z, z^{-1}] \quad \mapsto \quad \mathcal{I} \cap \mathbb{F}[z].$$

It preserves inclusions and takes prime ideals to prime ideals [18, p. 61]. Thus we can reduce the investigation of Laurent polynomial ideals, e.g., the computation of their dimension, to the study of ordinary polynomial ideals.

With these generalizations, a behavior will be a subset of $(\mathbb{F}^q)^T$, where in the most general case $T = \mathbb{Z}^{r_1} \times \mathbb{N}^{r_2}$, q is the number of components (e.g., inputs plus outputs) and $\mathbb{F}$ is some field (normally $\mathbb{R}$ or $\mathbb{C}$). A trajectory $w \in (\mathbb{F}^q)^T$ is a multi-indexed sequence taking its values in $\mathbb{F}^q$. We refer to the set T of multi-indices as the signal domain, and we always write $r = r_1 + r_2$ (in much existing work either r_1 or r_2 equals zero). Thus a behavior takes the form

$$\mathcal{B} = \ker R(\sigma) = \ker R(\sigma_1, \dots, \sigma_r) = \{w \in (\mathbb{F}^q)^T, \ R(\sigma)w = 0\}.$$

The matrix R is called a kernel representation matrix of $\mathcal{B}$. The module generated by its rows determines and is uniquely determined by $\mathcal{B}$, i.e., it does not depend on the particular choice of the representation matrix [48, p. 36]. Similarly, if

$$\mathcal{B} = \operatorname{im} M(\sigma) = \{w \in (\mathbb{F}^q)^T, \ \exists l \in (\mathbb{F}^m)^T \text{ such that } w = M(\sigma)l\}$$

for some polynomial $q \times m$ matrix M, we say that $\mathcal{B}$ has an image representation with image representation matrix M.

The property of having an image representation is important for many classes of systems [53, 59, 81, 84, 90] and it is equivalent to several other interesting properties. In particular, a behavior has an image representation iff it is minimal in its transfer class (combine [48, p. 142] with [84] or [90]).

Two behaviors are said to be **transfer equivalent** if the rows of their kernel representation matrices have the same span over the field of rational functions. The class $[\mathcal{B}]$ of all behaviors that are transfer equivalent to $\mathcal{B}$ is called the **transfer class** of $\mathcal{B}$. There exists a unique minimal element $\mathcal{B}^c \subseteq \mathcal{B}$ in $[\mathcal{B}]$, and $\mathcal{B}^c$ has an image representation. We say that $\mathcal{B}$ itself is **minimal** in its transfer class if $\mathcal{B} = \mathcal{B}^c$. The behavior $\mathcal{B}^c$ is precisely the "controllable part" of $\mathcal{B}$ discussed earlier, but it is only now that we give a justification of that nomenclature.

For systems defined on $\mathbb{Z}$ [81] or $\mathbb{Z}^2$ [59], the existence of an image representation has also been characterized in terms of **concatenability** of system trajectories that are specified on subsets of the signal domain that are "sufficiently far apart". This is a natural concept of controllability in the behavioral setting.

As we have indicated, there are some subtleties in the original 1D definition of controllability [81], which reveal themselves upon applying the definition to the signal domain $\mathbb{N}$. Due to these fine points, we will present several versions of Willems' original definition. Throughout this section, different definitions of controllability will be distinguished by indices.

Definition 7 *A 1D behavior $\mathcal{B}$ (defined on $T = \mathbb{Z}$) is said to be **controllable(1)** if there exists $\rho \in \mathbb{N}$ such that for all $w^{(1)}, w^{(2)} \in \mathcal{B}$, there exists $w \in \mathcal{B}$ such that*

$$w(t) = \begin{cases} w^{(1)}(t) & \text{if } t < 0 \\ w^{(2)}(t - \rho) & \text{if } t \geq \rho. \end{cases} \qquad (1.21)$$

The original definition of Willems [81] allows that length ρ of the transition period depends on the trajectories $w^{(1)}, w^{(2)}$ to be concatenated. For linear time-invariant behaviors however, ρ can be chosen uniformly for all trajectories.

Controllability(1) is obviously unsuitable for systems defined on $\mathbb{N}$, since it allows specification of the first trajectory only for negative times. We can get around this problem by moving the region of transition, *i.e.*, the time interval $[0, \rho]$, to an arbitrary location:

Definition 8 *A 1D behavior $\mathcal{B}$ (defined on $T = \mathbb{Z}$ or $T = \mathbb{N}$) is said to be **controllable(2)** if there exists $\rho \in \mathbb{N}$ such that, for any $w^{(1)}, w^{(2)} \in \mathcal{B}$, and any $t_0 \in T$, there exists $w \in \mathcal{B}$ such that*

$$w(t) = \begin{cases} w^{(1)}(t) & \text{if } t < t_0 \\ w^{(2)}(t - (t_0 + \rho)) & \text{if } t \geq t_0 + \rho. \end{cases} \qquad (1.22)$$

Clearly controllability(2) is equivalent to controllability(1) for systems with $T = \mathbb{Z}$. In generalizing controllability(2) to $T = \mathbb{Z}^2$ and thence to $T = \mathbb{Z}^r$, Rocha [59] observed that for these systems the shift of $w^{(2)}$ in equation (1.22) is not important. The existing rD definition [59, 84] requires an arbitrary metric, but we will find it convenient to introduce the following specific one:

$$d(T_1, T_2) = \min\{|t_1 - t_2|, \ t_1 \in T_1, t_2 \in T_2\},$$

where $|a| = \sum_{i=1}^{r} |a_i|$ for $a \in \mathbb{Z}^r$.

Definition 9 *An rD behavior B (defined on $T = \mathbb{Z}^{r_1} \times \mathbb{N}^{r_2}$) is said to be* **controllable(3)** *if there exists $\rho \geq 0$ such that, for any $w^{(1)}, w^{(2)} \in B$, and any regions $T_1, T_2 \subset T$ with $d(T_1, T_2) > \rho$, there exists $w \in B$ such that*

$$w(t) = \begin{cases} w^{(1)}(t) & \text{if } t \in T_1 \\ w^{(2)}(t) & \text{if } t \in T_2. \end{cases} \tag{1.23}$$

Controllability(3) is equivalent to the earlier definitions for $T = \mathbb{Z}$. In the case $T = \mathbb{Z}^r$, we can prove [84]:

Theorem 6 *A behavior B with signal domain $\mathbb{Z}^r$ is controllable(3) if and only if it has an image representation.*

To see that controllability(3) is inappropriate for systems defined on $T = \mathbb{N}^r$, we need again the notion of autonomy. A set of free variables (inputs) [84, 90] of a behavior $B \subseteq (\mathbb{F}^q)^T$ is a set of components of w which are collectively unrestricted by the system laws $R(\sigma)w = 0$. The maximum size of such a set is called the number of free variables of B, and it has been shown [48, p. 38] to equal $q - \text{rank}(R)$, where R is an arbitrary kernel representation matrix of B. An autonomous behavior is one that is devoid of free variables, or equivalently, one whose kernel representation matrices have full column rank (compare Lemma 1.3.2).

Example 1.4.1. Consider the following 1D examples:

$$\begin{aligned} B_1 &= \{w \in \mathbb{R}^{\mathbb{N}}, \ w(t+1) = 0 \text{ for all } t \in \mathbb{N}\} \\ &= \ker R_1(\sigma), \quad R_1 = [z]. \\ B_2 &= \{w \in (\mathbb{R}^2)^{\mathbb{N}}, \ w_1(t+1) + w_1(t) = w_2(t), \\ & \qquad\qquad w_1(t) = w_2(t) - w_2(t+1) \quad \text{for all } t \in \mathbb{N}\} \\ &= \ker R_2(\sigma), \quad R_2 = \begin{bmatrix} z+1 & -1 \\ 1 & z-1 \end{bmatrix}. \end{aligned}$$

$$\begin{aligned} B_3 &= \{w \in (\mathbb{R}^2)^{\mathbb{N}}, \ w_1(t+3) - w_1(t+2) + w_2(t+4) + w_2(t+3) = 0 \\ & \qquad \text{for all } t \in \mathbb{N}\} \\ &= \ker R_3(\sigma), \quad R_3 = \begin{bmatrix} z^2(z-1) & z^3(z+1) \end{bmatrix}. \\ B_4 &= \{w \in (\mathbb{R}^3)^{\mathbb{N}}, \ w_1(t) = w_2(t), \ w_2(t+1) = w_3(t+2) \text{ for all } t \in \mathbb{N}\} \\ &= \ker R_4(\sigma), \quad R_4 = \begin{bmatrix} 1 & -1 & 0 \\ 0 & -z & z^2 \end{bmatrix}. \end{aligned}$$

$\mathcal{B}_1$ is zero everywhere except at time $t = 0$, when it can take any value. $\mathcal{B}_2$ is also zero except at $t = 0, 1$. These two behaviors are obviously autonomous, but are also controllable(3), since in this example the concatenation conditions of controllability(3) are trivial. Note also that both representations can be regarded as being in classical state-space form $w(t + 1) = Aw(t)$, where $A = 0$ in the first example and

$$A = \begin{bmatrix} -1 & 1 \\ -1 & 1 \end{bmatrix}$$

in the second. As systems without inputs, they are certainly not state-point controllable in the classical sense. We see furthermore that neither $\mathcal{B}_1$ nor $\mathcal{B}_2$ is minimal in its transfer class (the trivial behavior $\{0\}$ is the minimal element in each case), so the characterization of Theorem 6 must fail for $T = \mathbb{N}^r$.

The behaviors $\mathcal{B}_3$ and $\mathcal{B}_4$ are also controllable(3), which can be seen taking separation distances of $\rho = \frac{5}{2}, \rho = \frac{3}{2}$ respectively. However they do not satisfy the conditions of controllability(4) to be defined below, and therefore as we will see they admit no image representations (alternatively, we can argue that R_3 and R_4 are not left prime).

The existence in particular of autonomous controllable non-trivial behaviors is counter-intuitive, and suggests a problem with the definition of controllability. Trajectories in a system with signal domain $\mathbb{N}^r$ can behave in a different way close to the origin than arbitrarily far from it, and the controllability(3) definition does not require that the concatenating trajectory w exhibits this close-to-the-origin behavior of the trajectory $w^{(2)}$. Therefore, if the control problem requires the reproduction of an entire signal $w^{(2)}$, the controllability(3) condition will be insufficient. Essentially, this complication comes from the fact that σ_i is not an invertible operator on $(\mathbb{F}^q)^T$ for $T = \mathbb{N}^r$, unlike in the $\mathbb{Z}^r$ case. The definition of controllability needs to be adapted to take account of this.

1.4.2 A New Definition of Controllability

We will shortly present our new definition of behavioral controllability. This requires some preliminary notation. We will find it convenient to define the following obvious action of a shift on any subset T_1 of T:

$$\sigma^a T_1 := (-a + T_1) \cap T = \{t \in T, \, t + a \in T_1\}.$$

The **diameter** of a bounded set $T_1 \subset T$ is

$$\rho(T_1) = \max\{|t - t'|, \, t, t' \in T_1\}.$$

Finally, given a polynomial matrix M, its **support** is defined as follows: For $M = \sum_{a \in \mathbb{N}^r} M_a z^a$, with coefficient matrices M_a over $\mathbb{F}$,

$$\text{supp}(M) = \{a \in \mathbb{N}^r, \, M_a \neq 0\}.$$

Definition 10 *Let $\mathcal{B}$ be an rD behavior with signal domain $T = \mathbb{Z}^{r_1} \times \mathbb{N}^{r_2}$. Then $\mathcal{B}$ is said to be* **controllable(4)** *if there exists $\rho \geq 0$ such that for all $T_1, T_2 \subset T$ with $d(T_1, T_2) > \rho$, and for all $w^{(1)}, w^{(2)} \in \mathcal{B}$, and all $b_1, b_2 \in T$, there exists $w \in \mathcal{B}$ such that*

$$\sigma^{b_1} w = w^{(1)} \text{ on } \sigma^{b_1} T_1 \quad \text{and} \quad \sigma^{b_2} w = w^{(2)} \text{ on } \sigma^{b_2} T_2, \tag{1.24}$$

i.e.,
$$w(t) = \begin{cases} w^{(1)}(t - b_1) & \text{if } t \in T_1 \text{ and } t - b_1 \in T \\ w^{(2)}(t - b_2) & \text{if } t \in T_2 \text{ and } t - b_2 \in T. \end{cases} \tag{1.25}$$

In that case, we also say that $\mathcal{B}$ is controllable(4) with **separation distance** *ρ.*

The definitions of controllability which we have already presented are easily seen to be special cases of controllability(4). In particular, controllability(3), which works perfectly well for $T = \mathbb{Z}^r$, is seen to be controllability(4) for $b_1 = b_2 = 0$ (and it is easy to show that these definitions are equivalent for such T). For $T = \mathbb{N}$, controllability(4) is in fact equivalent to controllability(2).

As commented above, to derive previous characterizations of controllability in the case $T = \mathbb{Z}^r$ for the general case, it is sufficient to prove the following:

Theorem 7 *Let $\mathcal{B}$ be an rD behavior with signal domain $T = \mathbb{Z}^{r_1} \times \mathbb{N}^{r_2}$. The following are equivalent:*

1. *$\mathcal{B}$ is controllable(4);*
2. *$\mathcal{B}$ has an image representation.*

Moreover, if M is an image representation matrix for $\mathcal{B}$, then $\mathcal{B}$ is controllable(4) with separation distance $\rho(\mathrm{supp}(M))$.

Proof. "1 $\Rightarrow$ 2": Let $\mathcal{B}^c$ denote the minimal element of the transfer class of $\mathcal{B}$. Then $\mathcal{B}^c$ has an image representation. Let R and R^c be kernel representation matrices of $\mathcal{B}$ and $\mathcal{B}^c$, respectively; then there is a rational function matrix X such that $R^c = XR$. Write $X = N/d$ with a polynomial matrix N and a polynomial d.

Suppose that $\mathcal{B}$ is controllable(4) with separation distance ρ. We will prove that $\mathcal{B} = \mathcal{B}^c$, and hence that $\mathcal{B}$ has an image representation, as required. If $\mathcal{B}^c = (\mathbb{F}^q)^T$ (that is, $R^c = 0$), or if d is a constant polynomial, this is trivial, so assume otherwise. Let $w^{(1)}$ denote the zero trajectory and let $w^{(2)}$ be an arbitrary trajectory of $\mathcal{B}$. We will prove that $w^{(2)} \in \mathcal{B}^c$.

Next, let $\deg(d)$ denote the exponent in $\mathbb{N}^r$ corresponding to the head term of d (with respect to an arbitrary term order), and define:

$$T_1 = \{a + s, \quad a \in \mathrm{supp}(R^c), \ s \in \mathbb{N}^r \setminus (\deg(d) + \mathbb{N}^r)\} \tag{1.26}$$
$$T_2 = b_2 + \mathbb{N}^r, \tag{1.27}$$

where b_2 is chosen such that the distance between T_1 and T_2 is greater than ρ. Although T_1 and T_2 are contained in $\mathbb{N}^r$, we consider them as subsets of T. Finally, let $b_1 = 0$. Apply the new definition of controllability; let $w \in \mathcal{B}$ be a connecting trajectory. Note that $R(\sigma)w = 0$, and so $(dR^c)(\sigma)w = (NR)(\sigma)w = 0$.

Let $R^c(z) = \sum_a R_a^c z^a$, where the summation runs over all $a \in \text{supp}(R^c)$. As $w = w^{(1)}$ on T_1 we have that $(R^c(\sigma)w)(s) = \sum R_a^c w(a + s) = 0$ for all $s \in \mathbb{N}^r \setminus (\deg(d) + \mathbb{N}^r)$. But $d(\sigma)(R^c(\sigma)w) = 0$, hence $R^c(\sigma)w = 0$ on the whole of $\mathbb{N}^r \subseteq T$. This is due to the fact that, for any monomial ordering, a solution v of $(d(\sigma)v)(t) = 0$ for all $t \in \mathbb{N}^n$, is uniquely determined by the values of v on the set $\mathbb{N}^r \setminus (\deg(d) + \mathbb{N}^r)$.

Since $\sigma^{b_2} T_2 = \mathbb{N}^r$, $w^{(2)}$ is equal to a shift σ^{b_2} of w on all of $\mathbb{N}^r$, and so $(R^c w^{(2)})(t) = 0$ for all $t \in \mathbb{N}^r$. This argument can be re-applied in each hyperquadrant of $\mathbb{Z}^{r_1} \times \mathbb{N}^{r_2}$, for appropriate choices of T_1, T_2 and hence w, and so $R^c w^{(2)}$ vanishes on the whole of T. Therefore $w^{(2)} \in \mathcal{B}^c$, and so $\mathcal{B} = \mathcal{B}^c$.

"2 $\Rightarrow$ 1": Suppose that M is an image representation matrix of $\mathcal{B}$. We will prove that $\mathcal{B}$ is controllable(4) with separation distance $\rho := \rho(\text{supp}(M))$, thus establishing the final claim also. This part of the proof follows the lines of Rocha's original proof [59]. Let $w^{(1)} = M(\sigma)l^{(1)}, w^{(2)} = M(\sigma)l^{(2)}$ be given, and let $b_1, b_2 \in T$ be arbitrary. Let $T_1, T_2 \subseteq T$ be such that $d(T_1, T_2) > \rho$. Now for any $t_1 \in T_1, t_2 \in T_2$ and any $s, s' \in \text{supp}(M)$, we must have

$$|t_1 - t_2| \geq d(T_1, T_2) > \rho = \rho(\text{supp}(M)) \geq |s - s'|.$$

This yields that

$$(T_1 + \text{supp}(M)) \cap (T_2 + \text{supp}(M)) = \emptyset,$$

and now we see that the following is well-defined:

$$l(a) = \begin{cases} l^{(1)}(a - b_1) & \text{if } a \in T_1 + \text{supp}(M) \text{ and } a - b_1 \in T \\ l^{(2)}(a - b_2) & \text{if } a \in T_2 + \text{supp}(M) \text{ and } a - b_2 \in T \\ 0 & \text{otherwise.} \end{cases}$$

Then $w := M(\sigma)l \in \mathcal{B}$ and

$$w(s) = (M(\sigma)l)(s) = \sum_{a \in \text{supp}(M)} M_a l(a + s).$$

For $t \in \sigma^{b_1} T_1$, say $t = s - b_1$, $s \in T_1$, we have

$$(\sigma^{b_1} w)(t) = w(s) = \sum_{a \in \text{supp}(M)} M_a l(a + s)$$

$$= \sum_{a \in \text{supp}(M)} M_a l^{(1)}(a + s - b_1) = w^{(1)}(s - b_1) = w^{(1)}(t),$$

so $\sigma^{b_1} w = w^{(1)}$ on $\sigma^{b_1} T_1$, and similarly $\sigma^{b_2} w = w^{(2)}$ on $\sigma^{b_2} T_2$. Thus $\mathcal{B}$ is controllable(4). $\qquad\square$

Combining the new result of Theorem 7 with existing characterizations of behaviors with image representations [48, 84, 90], we obtain a further corollary; this requires a preliminary definition. A polynomial matrix R is **generalized factor left prime (GFLP) with respect to a ring** $\mathcal{D}$, where

$$\mathbb{F}[z] \subseteq \mathcal{D} \subseteq \mathbb{F}[z, z^{-1}]$$

if the existence of a factorization $R = DR_1$ with polynomial matrices D, R_1 and rank (R) = rank (R_1), implies that there exists a $\mathcal{D}$-matrix E such that $R_1 = ER$.

Corollary 3 *The following are equivalent:*

1. $\mathcal{B}$ is controllable(4);
2. $\mathcal{B}$ has an image representation;
3. $\mathcal{B}$ is minimal in its transfer class, i.e., $\mathcal{B} = \mathcal{B}^c$;
4. Any kernel representation matrix R of $\mathcal{B}$ is GFLP with respect to the ring $\mathcal{D}$ that corresponds to the signal domain of $\mathcal{B}$, that is

$$\mathcal{D} = \mathbb{F}[z_1, \ldots, z_{r_1}, z_{r_1+1}, \ldots, z_r, z_1^{-1}, \ldots, z_{r_1}^{-1}]$$

   for $T = \mathbb{Z}^{r_1} \times \mathbb{N}^{r_2}$;
5. There is no proper sub-behavior of $\mathcal{B}$ with the same number of free variables;
6. $\mathcal{B}$ is divisible, i.e., for any $p \in \mathcal{D} \setminus \{0\}$ it must hold that $\mathcal{B} = p(\sigma)\mathcal{B}$.

Moreover, if $\mathcal{B}$ is controllable(4) and autonomous, then $\mathcal{B} = \{0\}$.

This result applies to all standard classes of discrete systems. For completeness we mention that Pillai and Shankar [53] have established the equivalence of 1 and 2 for continuous systems, and in fact this is sufficient to establish Corollary 3 for all classes of systems considered in [48].

An obvious question to consider now is the relationship between controllability(3) and controllability(4). In fact, we can characterize this, which requires the following definition:

Definition 11 *An rD behavior $\mathcal{B}$ is said to be **permanent** if $\sigma_i \mathcal{B} = \mathcal{B}$ for $i = 1, \ldots, r$.*

Permanence coincides with shift-invariance in the case of behaviors defined on $\mathbb{Z}^r$, and is therefore trivial for behaviors with that signal domain. For behaviors over $\mathbb{N}^r$ however, shift-invariance implies only $\sigma_i \mathcal{B} \subseteq \mathcal{B}$, and permanence is strictly stronger than that. For a 1D behavior $\mathcal{B}$ over $\mathbb{N}$, with kernel representation matrix R, we mention without proof that permanence is equivalent to the condition that R does not lose rank at the origin.

Lemma 1.4.1. *An rD behavior $\mathcal{B}$ is controllable(4) iff it is controllable(3) and permanent.*

Proof. "if": Let $\mathcal{B}$ be controllable(3) with separation distance ρ. Let $T_1, T_2 \subset T$ be such that $d(T_1, T_2) > \rho$ and let $w^{(1)}, w^{(2)} \in \mathcal{B}$ and $b_1, b_2 \in T$ be given. By permanence, there exists $v^{(i)} \in \mathcal{B}$ such that $w^{(i)} = \sigma^{b_i} v^{(i)}$ for $i = 1, 2$. Let $w \in \mathcal{B}$ be the connecting trajectory of $v^{(1)}$, $v^{(2)}$ with respect to T_1, T_2 according to controllability(3). It is now easy to see that w is also the desired connecting trajectory of $w^{(1)}, w^{(2)}$ with respect to T_1, T_2 and b_1, b_2.

"only if": Controllability(4) obviously implies controllability(3) ($b_1 = b_2 = 0$) and, by Condition 6 of Corollary 7, we have that controllability(4) implies $\mathcal{B} = \sigma_i \mathcal{B}$ for all i.   □

The result of Lemma 1.4.1 should be compared to Staiger's [70] characterization of "remergeability," as discussed in [64]. Given this relationship between the two types of rD controllability, it is now natural to ask whether controllability(3) admits an algebraic or polynomial matrix characterization. Controllability(4) with $T = \mathbb{N}^r$ is equivalent to the condition that R is GFLP over $\mathbb{F}[z]$. Given Lemma 1.4.1, it is tempting to conjecture that the weaker notion of controllability(3) is equivalent to R being GFLP over $\mathbb{F}[z, z^{-1}]$. This is supported by the examples of the previous section, where the representing matrices are GFLP over $\mathbb{F}[z, z^{-1}]$, but not over $\mathbb{F}[z]$. But although controllability(3) does indeed imply generalized left factor primeness over the Laurent polynomial ring (see Lemma 1.4.2 below), the converse is not true in general:

Example 1.4.2. Consider the 2D behavior

$$\mathcal{B} = \{w \in \mathbb{R}^{\mathbb{N}^2}, \; w(t_1 + 1, t_2) = 0, \; w(t_1, t_2 + 1) = w(t_1, t_2) \;\; \forall \, t_1, t_2 \in \mathbb{N}\}$$

$$= \ker R(\sigma_1, \sigma_2), \quad R = \begin{bmatrix} z_1 \\ z_2 - 1 \end{bmatrix}.$$

Its trajectories are constant along the t_2-axis, and zero everywhere else. Thus $\mathcal{B}$ is not controllable(3) although R exhibits the property of being GFLP over

$$\mathbb{R}[z_1, z_2, z_1^{-1}, z_2^{-1}],$$

but not over $\mathbb{R}[z_1, z_2]$.

Thus controllability(3) of a behavior with signal domain $T = \mathbb{N}^r$ corresponds to a property of its kernel representation matrices that is weaker than "GFLP over $\mathbb{F}[z]$" and stronger than "GFLP over $\mathbb{F}[z, z^{-1}]$". This supports our suggestion that controllability(3) is not appropriate for systems defined on $\mathbb{N}^r$ at all.

Lemma 1.4.2. *Let $\mathcal{B}$ be an rD behavior with signal domain $T = \mathbb{Z}^{r_1} \times \mathbb{N}^{r_2}$. If $\mathcal{B} = \ker R(\sigma)$ is controllable(3), then R is GFLP with respect to $\mathbb{F}[z, z^{-1}]$.*

Proof. Let R^c be a kernel representation matrix of $\mathcal{B}^c$. Then R^c is GFLP over $\mathbb{F}[z]$ (and thus over $\mathbb{F}[z, z^{-1}]$). Let g and g' denote the number of rows of

R and R^c, respectively. A variant of the proof "$1 \Rightarrow 2$" in Theorem 7 shows that if $\mathcal{B}$ is controllable(3), then there exists $b \in \mathbf{N}^r$ such that $\sigma^b \mathcal{B} \subseteq \mathcal{B}^c$, that is, $w \in \mathcal{B} \Rightarrow \sigma^b w \in \mathcal{B}^c \subseteq \mathcal{B}$. It follows that $\ker R^c(\sigma) \subseteq \ker R(\sigma) \subseteq \ker \sigma^b R^c(\sigma)$, or

$$\mathbb{F}[z]^{1 \times g'} z^b R^c \subseteq \mathbb{F}[z]^{1 \times g} R \subseteq \mathbb{F}[z]^{1 \times g'} R^c.$$

This implies that the rows of R and R^c generate the same module over $\mathbb{F}[z, z^{-1}]$, and so the behaviors on $\mathbb{Z}^r$ represented by R and R^c coincide. Since R^c is GFLP over $\mathbb{F}[z, z^{-1}]$, R must also be. □

In the proof of Theorem 7, we used controllability(4) only with $b_1 = 0$ in order to prove that $\mathcal{B}$ had an image representation. Thus controllability(4) is equivalent to the same property but with the restriction $b_1 = 0$. For the signal domain $T = \mathbb{Z}^r$, we can take $b_2 = 0$ as well, returning us to controllability(3). This is of course not possible for $T = \mathbf{N}^r$, though in this case a different choice of b_2 or of b_1 characterizing controllability(4) is possible, as we see following the next definition.

Definition 12 *For $\emptyset \neq T_1 \subseteq \mathbf{N}^r$, let the **cw-infimum** of T_1, denoted by $\kappa(T_1)$, be defined as the minimal element of the smallest (possibly infinite) interval of $\mathbf{N}^r$ containing T_1.*

Corollary 4 *Let $T = \mathbf{N}^r$. The behavior $\mathcal{B}$ is controllable(4) iff the following two equivalent conditions are satisfied:*

(a) There exists $\rho \geq 0$ such that for all $T_1, T_2 \subset \mathbf{N}^r$ with $d(T_1, T_2) > \rho$ and for all $w^{(1)}, w^{(2)} \in \mathcal{B}$, there exists $w \in \mathcal{B}$ such that

$$\sigma^{\kappa(T_1)} w = w^{(1)} \text{ on } \sigma^{\kappa(T_1)} T_1 \quad \text{and} \quad \sigma^{\kappa(T_2)} w = w^{(2)} \text{ on } \sigma^{\kappa(T_2)} T_2,$$

$$\text{i.e.,} \qquad w(t) = \begin{cases} w^{(1)}(t - \kappa(T_1)) & \text{if } t \in T_1 \\ w^{(2)}(t - \kappa(T_2)) & \text{if } t \in T_2. \end{cases}$$

(b) There exists $\rho \geq 0$ such that for all $T_1, T_2 \subset \mathbf{N}^r$ with $d(T_1, T_2) > \rho$ and for all $w^{(1)}, w^{(2)} \in \mathcal{B}$, there exists $w \in \mathcal{B}$ such that

$$w = w^{(1)} \text{ on } T_1 \quad \text{and} \quad \sigma^{\kappa(T_2)} w = w^{(2)} \text{ on } \sigma^{\kappa(T_2)} T_2,$$

$$\text{i.e.,} \qquad w(t) = \begin{cases} w^{(1)}(t) & \text{if } t \in T_1 \\ w^{(2)}(t - \kappa(T_2)) & \text{if } t \in T_2. \end{cases}$$

Proof. Both conditions are special cases of the condition controllability(4). Conversely, if a behavior $\mathcal{B}$ is controllable(4), then by Theorem 7 it has an image representation. By applying a variant of the proof "$2 \Rightarrow 1$" of Theorem 7, we can easily establish that $\mathcal{B}$ satisfies (a) or (b). Note that for T_1, T_2 defined according to (1.26), (1.27), we have $\kappa(T_1) = b_1 = 0$ and $\kappa(T_2) = b_2$. □

Note how condition (b) coincides with controllability(2) in the 1D case $(T = \mathbb{N})$ when we consider "past" $T_1 = \{0, \ldots, t_0 - 1\}$ and "future" $T_2 = \{t_0 + \rho, \ldots\}$. Clearly, $d(T_1, T_2) > \rho$ and $\kappa(T_2) = t_0 + \rho$.

1.4.3 Behavioral and Classical Definitions

Following Willems [81], we now describe the relationship between behavioral controllability and "state-point controllability", *i.e.*, classical controllability (or reachability) in terms of the state space.

Let $\mathcal{B}_{s,i}$ denote the state-input behavior of a 1D classical state-space representation, *i.e.*,

$$\mathcal{B}_{s,i} = \ker R(\sigma) \quad \text{with} \quad R = [\ zI - A, \ -B\ ],$$

where A and B are matrices over $\mathbb{F}$. In other words, $\mathcal{B}_{s,i}$ contains all pairs (x, u) that satisfy $x(t + 1) = Ax(t) + Bu(t)$ for all $t \in T$. The behavior $\mathcal{B}_{s,i}$ is said to be **state trim** if any state x_0 occurs as an initial state in the sense that there exists $(x, u) \in \mathcal{B}_{s,i}$ such that $x(0) = x_0$. For $T = \mathbb{N}$, $\mathcal{B}_{s,i}$ is always state trim, since for any state x_0 we can simply set $x(0) := x_0$ and allow the trajectory $w = (x, u)$ to evolve according to the system equations.

Lemma 1.4.3.

1. *For $T = \mathbb{Z}$, (A, B) is state-point controllable iff $\mathcal{B}_{s,i}$ is controllable(4) and state trim.*
2. *For $T = \mathbb{N}$, (A, B) is state-point controllable iff $\mathcal{B}_{s,i}$ is controllable(4).*

Proof. The proof is essentially due to Willems [81]. For ease of exposition we consider only the cases $\mathbb{F} = \mathbb{R}, \mathbb{C}$. Other fields can be dealt with in a similar way, or by using an alternative trajectory proof.

In view of Theorem 7, we only need that – in the 1D case – $\mathcal{B}_{s,i}$ has an image representation iff R is left prime. In the case $T = \mathbb{N}$, this signifies that

$$R(\lambda) = [\ \lambda I - A, \ -B\ ]$$

has full row rank for all $\lambda \in \mathbb{C}$, which is the classical Hautus condition for controllability of the matrix pair (A, B). For $T = \mathbb{Z}$, left primeness of R is equivalent to $R(\lambda)$ having full row rank for all $0 \neq \lambda \in \mathbb{C}$. So we need the additional requirement "$[\ A, \ B\ ]$ has full row rank," corresponding to $\lambda = 0$. But this is precisely the condition for $\mathcal{B}_{s,i}$ to be state trim [81].   □

For the case $T = \mathbb{N}$, it is easy to see that Lemma 1.4.3 does not apply to controllability(3). We can take the simple earlier example $\mathcal{B} = \{w, w(t+1) = 0, t \geq 0\}$, treated as a state-space representation with no inputs and $A = 0$; this behavior is controllable(3) but certainly not state-point controllable.

1.5 Controllability as Parameterizability

The problem of characterizing the linear shift-invariant multidimensional behaviors that possess an image representation, is equivalent to giving an algebraic criterion for a system of linear, constant-coefficient partial differential equations to be "parameterizable," that is, derivable from a potential, as in

$$\text{curl}(w) = 0 \quad \Leftrightarrow \quad w = \text{grad}(\phi).$$

The aim of this section is to provide some interesting examples of systems of partial differential equations that admit such parameterizations, e.g., the Maxwell equations, and to investigate the homogeneous systems of linearized Ricci, Einstein, and Riemann equations with regard to parameterizability in that sense.

1.5.1 The Parameterizability Criterion

Consider the solution space of a linear system of partial differential equations with constant coefficients,

$$\mathcal{B} = \{w \in \mathcal{A}^q, R(\partial_1, \ldots, \partial_r)w = 0\}, \qquad (1.28)$$

where $\mathcal{A} = C^\infty(\mathbb{R}^r)$, the space of infinitely often differentiable functions $\mathbb{R}^r \to \mathbb{C}$, or $\mathcal{A} = \mathcal{D}'(\mathbb{R}^r)$, the space of distributions on $\mathbb{R}^r$. In both cases, our signal domain is the continuous set $\mathbb{R}^r$, and the signal space is the complex vector space $\mathcal{A}^q$, thus $\mathbb{F} = \mathbb{C}$. As usual, the kernel representation matrix R is a polynomial in r indeterminates s_i corresponding to the partial differential operators ∂_i,

$$R \in \mathbb{C}[s]^{g \times q}, \quad \mathbb{C}[s] := \mathbb{C}[s_1, \ldots, s_r].$$

We know that $\mathcal{B}$ admits a kernel representation (1.28) with a GFLP matrix R iff there exists $k \geq 1$ and $M \in \mathbb{C}[s]^{q \times k}$ such that the sequence

$$\mathbb{C}[s]^g \xrightarrow{R^T} \mathbb{C}[s]^q \xrightarrow{M^T} \mathbb{C}[s]^k \qquad (1.29)$$

is exact. In other words, R is a minimal left annihilator (MLA) of M, that is, its rows generate $\{\eta \in \mathbb{C}[s]^{1 \times q}, \eta M = 0\}$. But then, by a variant of Palamodov's fundamental theorem [51, pp. 289–303], the inhomogeneous system of equations $M(\partial_1, \ldots, \partial_r)l = w$ is solvable iff $R(\partial_1, \ldots, \partial_r)w = 0$. Thus we can write

$$\mathcal{B} = \{w \in \mathcal{A}^q, \exists l : w = M(\partial_1, \ldots, \partial_r)l\}. \qquad (1.30)$$

Conversely, (1.28) and (1.30) imply exactness of (1.29) due to Oberst's duality theorem [48], which applies also to the continuous signal spaces considered here. For convenience, we restate the test for the GFLP property:

1. Compute a minimal right annihilator (MRA) of R, call it M. In other words, construct an exact sequence

$$\mathbb{C}[s]^k \xrightarrow{M} \mathbb{C}[s]^q \xrightarrow{R} \mathbb{C}[s]^g.$$

2. Compute an MLA of M, call it R^c. We have

$$\mathbb{C}[s]^g \xrightarrow{R^T} \mathbb{C}[s]^q \xrightarrow{M^T} \mathbb{C}[s]^k$$
$$\mathbb{C}[s]^{g_c} \xrightarrow{R^{cT}}$$

where the lower sequence is exact.

3. Check whether the rows of R and R^c generate the same module. If yes, R is GFLP; moreover, it is an MLA of M and $\mathcal{B}$ has an image representation.

Recall that this algorithm also constructs the desired parameterization (1.30) of $\mathcal{B}$ (if it exists).

Using different techniques, Pommaret and Quadrat [55] also derived tests for a system of linear PDEs to be parameterizable. Their method applies also to the variable coefficient case.

1.5.2 Examples

Divergence, Curl, and Gradient. Consider the set of divergence-less fields

$$\mathcal{B} = \{w = (w_1, w_2, w_3)^T, \ \mathrm{div}(w) = \nabla \cdot w = 0\}.$$

Its kernel representation is

$$R = \begin{bmatrix} s_1 & s_2 & s_3 \end{bmatrix}.$$

An MRA is given by

$$M = \begin{bmatrix} 0 & -s_3 & s_2 \\ s_3 & 0 & -s_1 \\ -s_2 & s_1 & 0 \end{bmatrix}.$$

Computing the first syzygy of the module generated by the rows of M yields

$$R^c = \begin{bmatrix} s_1 & s_2 & s_3 \end{bmatrix}.$$

Obviously, R and R^c coincide. We obtain the desired parameterization of $\mathcal{B}$:

$$\mathrm{div}(w) = 0 \quad \Leftrightarrow \quad \exists l : w = \mathrm{curl}(l).$$

The dual result concerns the set of conservative vector fields

$$\mathcal{B} = \{w = (w_1, w_2, w_3)^T, \ \mathrm{curl}(w) = \nabla \times w = 0\}.$$

Here,

$$R = \begin{bmatrix} 0 & -s_3 & s_2 \\ s_3 & 0 & -s_1 \\ -s_2 & s_1 & 0 \end{bmatrix}$$

and an MRA is

$$M = \begin{bmatrix} s_1 \\ s_2 \\ s_3 \end{bmatrix}.$$

But R is also an MLA of M, thus we obtain, as expected,

$$\text{curl}(w) = 0 \quad \Leftrightarrow \quad \exists \phi : w = \text{grad}(\phi) = \nabla \phi.$$

It should be noted that whenever one of the equations in $\text{curl}(w) = 0$ is dropped, the resulting system of PDEs is *not* parameterizable, although we know from a previous example that the corresponding kernel representation matrix, say

$$\tilde{R} = \begin{bmatrix} 0 & -s_3 & s_2 \\ s_3 & 0 & -s_1 \end{bmatrix}$$

is even factor left prime in the classical sense (that is, in any factorization $\tilde{R} = D\tilde{R}_1$, the square matrix D is necessarily unimodular).

Maxwell Equations. The first set of Maxwell equations is given by

$$\frac{\partial B}{\partial t} + \nabla \times E = 0 \tag{1.31}$$

$$\nabla \cdot B = 0. \tag{1.32}$$

Let $w = (B_1, B_2, B_3, E_1, E_2, E_3)^T$ be the vector of unknowns, and let

$$\mathcal{B} = \{w, \ w \text{ satisfies } (1.31), (1.32)\}.$$

Identifying $s_4 = \partial_t$, a kernel representation is

$$R = \begin{bmatrix} s_4 & 0 & 0 & 0 & -s_3 & s_2 \\ 0 & s_4 & 0 & s_3 & 0 & -s_1 \\ 0 & 0 & s_4 & -s_2 & s_1 & 0 \\ s_1 & s_2 & s_3 & 0 & 0 & 0 \end{bmatrix}.$$

Note that $\text{rank}(R) = 3$. First we compute a MRA of R,

$$M = \begin{bmatrix} 0 & -s_3 & s_2 & 0 \\ s_3 & 0 & -s_1 & 0 \\ -s_2 & s_1 & 0 & 0 \\ -s_4 & 0 & 0 & -s_1 \\ 0 & -s_4 & 0 & -s_2 \\ 0 & 0 & -s_4 & -s_3 \end{bmatrix}$$

and then we compute an MLA of M, which again turns out to be an identical copy of R itself. Thus, (1.31), (1.32) are parametrized by M, reproducing the well-known parameterization of $\boldsymbol{E}$ and $\boldsymbol{B}$ in terms of the magnetic vector potential $\boldsymbol{A}$ and the scalar electric potential ϕ

$$\boldsymbol{B} = \nabla \times \boldsymbol{A}$$
$$\boldsymbol{E} = -\frac{\partial \boldsymbol{A}}{\partial t} - \nabla \phi.$$

For the second set of Maxwell equations,

$$\nabla \times \boldsymbol{H} = \boldsymbol{J} + \frac{\partial \boldsymbol{D}}{\partial t} \tag{1.33}$$

$$\nabla \cdot \boldsymbol{D} = \rho, \tag{1.34}$$

let $w = (D_1, D_2, D_3, H_1, H_2, H_3, J_1, J_2, J_3, \rho)^T$, and a kernel representation is given by

$$R = \begin{bmatrix} s_4 & 0 & 0 & 0 & s_3 & -s_2 & 1 & 0 & 0 & 0 \\ 0 & s_4 & 0 & -s_3 & 0 & s_1 & 0 & 1 & 0 & 0 \\ 0 & 0 & s_4 & s_2 & -s_1 & 0 & 0 & 0 & 1 & 0 \\ s_1 & s_2 & s_3 & 0 & 0 & 0 & 0 & 0 & 0 & -1 \end{bmatrix}$$

Unlike the first set of Maxwell equations, this system of equations has full rank, in fact, it even possesses a polynomial right inverse, in other words, the matrix R is zero left prime. We compute its module of syzygies and obtain

$$M = \begin{bmatrix} 1 & 0 & 0 & 0 & 0 & 0 \\ 0 & 1 & 0 & 0 & 0 & 0 \\ 0 & 0 & 1 & 0 & 0 & 0 \\ 0 & 0 & 0 & 1 & 0 & 0 \\ 0 & 0 & 0 & 0 & 1 & 0 \\ 0 & 0 & 0 & 0 & 0 & 1 \\ -s_4 & 0 & 0 & 0 & -s_3 & s_2 \\ 0 & -s_4 & 0 & s_3 & 0 & -s_1 \\ 0 & 0 & -s_4 & -s_2 & s_1 & 0 \\ s_1 & s_2 & s_3 & 0 & 0 & 0 \end{bmatrix}$$

Here, the parameterization yields nothing new, it just implies that $\boldsymbol{D}$ and $\boldsymbol{H}$ are free, and reproduces the parameterization of $\boldsymbol{J}$ and ρ in terms of $\boldsymbol{D}, \boldsymbol{H}$ that is already contained in the equations (1.33), (1.34) themselves.

Ricci, Einstein, and Riemann Equations. The following problem was posed by Wheeler in 1970, and it was solved by Pommaret [56, 54] in 1995. Here we reproduce his result using our criterion. Wheeler's challenge was to find out whether or not the linearized Ricci equations in vacuum [8]

$$
\begin{aligned}
0 = \; & \partial_{ij}(\Omega_{11} + \Omega_{22} + \Omega_{33} - \Omega_{44}) \\
& + (\partial_{11} + \partial_{22} + \partial_{33} - \partial_{44})\Omega_{ij} \\
& - (\partial_{1i}\Omega_{1j} + \partial_{2i}\Omega_{2j} + \partial_{3i}\Omega_{3j} - \partial_{4i}\Omega_{4j}) \\
& - (\partial_{1j}\Omega_{1i} + \partial_{2j}\Omega_{2i} + \partial_{3j}\Omega_{3i} - \partial_{4j}\Omega_{4i})
\end{aligned}
\tag{1.35}
$$

where $1 \le i \le j \le 4$, admit a potential in the sense of the previous examples. Here, Ω denotes a symmetric tensor describing a small perturbation of the flat Minkowski metric $\mathrm{diag}(1,1,1,-1)$. In fact, the sign pattern in (1.35) is due to certain contractions with this metric.

The answer to Wheeler's question is negative, which can be seen setting

$$
w = (\Omega_{11}, \Omega_{22}, \Omega_{33}, \Omega_{44}, \Omega_{12}, \Omega_{23}, \Omega_{34}, \Omega_{13}, \Omega_{24}, \Omega_{14})^T
$$

and rewriting (1.35) with the kernel representation matrix

$$
R = \left[
\begin{array}{cccc}
s_2^2 + s_3^2 - s_4^2 & s_1^2 & s_1^2 & -s_1^2 \\
s_2^2 & s_1^2 + s_3^2 - s_4^2 & s_2^2 & -s_2^2 \\
s_3^2 & s_3^2 & s_1^2 + s_2^2 - s_4^2 & -s_3^2 \\
s_4^2 & s_4^2 & s_4^2 & s_1^2 + s_2^2 + s_3^2 \\
0 & 0 & s_1 s_2 & -s_1 s_2 \\
s_2 s_3 & 0 & 0 & -s_2 s_3 \\
s_3 s_4 & s_3 s_4 & 0 & 0 \\
0 & s_1 s_3 & 0 & -s_1 s_3 \\
s_2 s_4 & 0 & s_2 s_4 & 0 \\
0 & s_1 s_4 & s_1 s_4 & 0
\end{array}
\right.
$$

$$
\left.
\begin{array}{cccccc}
-2s_1 s_2 & 0 & 0 & -2s_1 s_3 & 0 & 2s_1 s_4 \\
-2s_1 s_2 & -2s_2 s_3 & 0 & 0 & 2s_2 s_4 & 0 \\
0 & -2s_2 s_3 & 2s_3 s_4 & -2s_1 s_3 & 0 & 0 \\
0 & 0 & -2s_3 s_4 & 0 & -2s_2 s_4 & -2s_1 s_4 \\
s_3^2 - s_4^2 & -s_1 s_3 & 0 & -s_2 s_3 & s_1 s_4 & s_2 s_4 \\
-s_1 s_3 & s_1^2 - s_4^2 & s_2 s_4 & -s_1 s_2 & s_3 s_4 & 0 \\
0 & -s_2 s_4 & s_1^2 + s_2^2 & -s_1 s_4 & -s_2 s_3 & -s_1 s_3 \\
-s_2 s_3 & -s_1 s_2 & s_1 s_4 & s_2^2 - s_4^2 & 0 & s_3 s_4 \\
-s_1 s_4 & -s_3 s_4 & -s_2 s_3 & 0 & s_1^2 + s_3^2 & -s_1 s_2 \\
-s_2 s_4 & 0 & -s_1 s_3 & -s_3 s_4 & -s_1 s_2 & s_2^2 + s_3^2
\end{array}
\right]
$$

An MRA of R is given by

$$M = \begin{bmatrix} 2s_1 & 0 & 0 & 0 \\ 0 & 2s_2 & 0 & 0 \\ 0 & 0 & 2s_3 & 0 \\ 0 & 0 & 0 & -2s_4 \\ s_2 & s_1 & 0 & 0 \\ 0 & s_3 & s_2 & 0 \\ 0 & 0 & s_4 & -s_3 \\ s_3 & 0 & s_1 & 0 \\ 0 & s_4 & 0 & -s_2 \\ s_4 & 0 & 0 & -s_1 \end{bmatrix} \tag{1.36}$$

Note that $w = M(\partial_1, \dots, \partial_4)l$ iff $\Omega_{ij} = \partial_i l_j + \partial_j l_i$, that is, iff Ω is the Lie derivative of the flat Minkowski metric along the vector field $l = (l_1, l_2, l_3, -l_4)$. It turns out that the syzygy of the module generated by the rows of M is strictly greater than the row module of R (in fact, the 20×10 matrix R^c corresponds precisely to the linearized Riemann operator, which is given in an appendix at the end of this section). This implies that equations (1.35) cannot be parametrized in the sense discussed above. For instance,

$$\begin{bmatrix} s_2^2 & s_1^2 & 0 & 0 & -2s_1 s_2 & 0 & 0 & 0 & 0 & 0 \end{bmatrix}$$

is in the syzygy of M, but the corresponding PDE

$$\partial_{22} \Omega_{11} + \partial_{11} \Omega_{22} - 2\partial_{12} \Omega_{12} = 0$$

cannot be deduced from (1.35) alone.

Finally, the Einstein tensor is the trace-reversed Ricci tensor. For the linearized Einstein equations, the first four rows of the matrix R from above have to be replaced by:

$$\begin{bmatrix} 0 & -s_3^2 + s_4^2 & -s_2^2 + s_4^2 & s_2^2 + s_3^2 \\ -s_3^2 + s_4^2 & 0 & -s_1^2 + s_4^2 & s_1^2 + s_3^2 \\ -s_2^2 + s_4^2 & -s_1^2 + s_4^2 & 0 & s_1^2 + s_2^2 \\ s_2^2 + s_3^2 & s_1^2 + s_3^2 & s_1^2 + s_2^2 & 0 \end{bmatrix}$$

$$\begin{bmatrix} 0 & 2s_2 s_3 & -2s_3 s_4 & 0 & -2s_2 s_4 & 0 \\ 0 & 0 & -2s_3 s_4 & 2s_1 s_3 & 0 & -2s_1 s_4 \\ 2s_1 s_2 & 0 & 0 & 0 & -2s_2 s_4 & -2s_1 s_4 \\ -2s_1 s_2 & -2s_2 s_3 & 0 & -2s_1 s_3 & 0 & 0 \end{bmatrix}$$

Pre-multiplying the remaining six rows by the unimodular matrix

$$U = \mathrm{diag}(2, 2, -2, 2, -2, -2)$$

yields a symmetric kernel representation R_1 of rank 6. The matrices R and R_1 have the same module of syzygies, that is, the matrix M from (1.36) above is an MRA also of R_1. It turns out that the linearized Einstein equations are not parameterizable for the same reasons as with the linearized Ricci equations. Therefore these systems of linear constant-coefficient PDEs do not admit a generic potential like the Maxwell equations.

As a by-product however, we get that the linearized Riemann equations (characterizing metrics that are constant in some coordinate system) can be parametrized by M from (1.36), that is, Ω is in the kernel of the linearized Riemann operator iff it is the Lie derivative of the flat Minkowski metric along some vector field.

Appendix

Linearized Riemann operator.

$$
\begin{bmatrix}
s_2^2 & & s_1^2 & & & -2s_1s_2 & & & & & \\
& & s_3^2 & s_2^2 & & & -2s_2s_3 & & & & \\
& & & s_4^2 & s_3^2 & & & -2s_3s_4 & & & \\
s_3^2 & & s_1^2 & & & & & & -2s_1s_3 & & \\
& s_4^2 & & s_2^2 & & & & & & -2s_2s_4 & \\
s_4^2 & & & s_1^2 & & & & & & & -2s_1s_4 \\
s_2s_3 & & & & & -s_1s_3 & s_1^2 & & -s_1s_2 & & \\
s_3s_4 & & & & & & & s_1^2 & -s_1s_4 & & -s_1s_3 \\
s_2s_4 & & & & & -s_1s_4 & & s_1^2 & & & -s_1s_2 \\
& s_3s_4 & & & & & -s_2s_4 & s_2^2 & & -s_2s_3 & \\
& s_1s_3 & & & & & -s_2s_3 & -s_1s_2 & s_2^2 & & \\
& s_1s_4 & & & & & -s_2s_4 & & & -s_1s_2 & s_2^2 \\
& & s_1s_2 & & & s_3^2 & -s_1s_3 & & & -s_2s_3 & \\
& & s_2s_4 & & & & -s_3s_4 & -s_2s_3 & & s_3^2 & \\
& & s_1s_3 & & & & & & -s_1s_3 & -s_3s_4 & s_3^2 \\
& & & s_1s_2 & s_4^2 & & & & & -s_1s_4 & -s_2s_4 \\
& & & s_2s_3 & & & s_4^2 & & -s_2s_4 & -s_3s_4 & \\
& & & s_1s_3 & & & & s_4^2 & -s_1s_4 & & -s_3s_4 \\
& & & & & -s_3s_4 & s_1s_4 & -s_1s_2 & & & s_2s_3 \\
& & & & & & s_1s_4 & & -s_2s_4 & -s_1s_3 & s_2s_3
\end{bmatrix}
$$

2. Co-prime Factorizations of Multivariate Rational Matrices

Co-prime factorization is a well-known issue in one-dimensional systems theory, having many applications in realization theory, balancing, controller synthesis, etc. Generalization to systems in more than one independent variable is a delicate matter: First of all, there are several non-equivalent co-primeness notions for multivariate polynomial matrices: zero, minor, and factor co-primeness. Here, we adopt the generalized version of factor primeness discussed in the previous chapter: A matrix is prime iff it is a minimal annihilator. After reformulating the sheer concept of a factorization, it is shown that every rational matrix possesses left and right co-prime factorizations that can be found by means of computer algebraic methods. Several properties of co-prime factorizations are given in terms of certain determinantal ideals.

Matrix fraction descriptions of rational matrices are prominent in algebraic systems theory, see for instance Kailath [35]. The idea is to express a rational matrix, usually interpreted as the transfer operator of a linear time-invariant system, as the "ratio" of two relatively prime polynomial matrices, just like in the scalar case, where any rational function can be represented as the quotient of two co-prime polynomials ("pole-zero cancellation" in systems theory's folklore).

Of course, commutativity is lost when passing from the scalar to the multi-variable case. Thus, for a rational transfer matrix H, left and right factorizations have to be distinguished:

$$H = D^{-1}N \quad \text{and} \quad H = \bar{N}\bar{D}^{-1},$$

the former corresponding to an input-output relation $Dy = Nu$, the latter to a driving-variable description $u = \bar{D}v, y = \bar{N}v$.

An irreducible matrix fraction description, or a co-prime factorization, is such that the numerator matrix N ($\bar{N}$) and the denominator matrix D ($\bar{D}$) are devoid of non-trivial common left (right) factors. The determinantal degree of such a (left or right) co-prime factorization is minimal among the determinantal degrees of all possible factorizations of a rational matrix. This fact is crucial for realization theory.

The transfer matrices of multidimensional systems do not depend on one single variable (usually interpreted as frequency), but on several independent variables. The theory is far less developed. This is partially due to the annoying (or: fascinating) diversity of multivariate primeness notions. Here, we stick to the notion of generalized factor primeness which has proven to be adequate for multidimensional systems (we tacitly mean the case when $r > 2$, the 2D case has its own distinctive features, see Section 2.6).

The chapter is organized as follows: After presenting the required concepts of multivariate factorizations in Section 2.1, co-primeness is characterized in Section 2.2, and a computational primeness test is given. The short but crucial Section 2.3 shows how to construct co-prime factorizations of arbitrary multivariate rational matrices. Section 2.4 discusses some of the most important properties of co-prime factorizations. Finally, Section 2.5 provides further insight into the structure of co-prime factorizations, based on the theory of factorization ideals. The final section treats the bivariate case.

2.1 Notions

As usual, let $\mathbb{F}$ be a field, and let $\mathcal{A}$ be one of the following signal spaces: $\mathcal{A} = C^\infty(\mathbb{R}^r)$ or $\mathcal{A} = \mathcal{D}'(\mathbb{R}^r)$ (with $\mathbb{F} = \mathbb{R}$ or $\mathbb{F} = \mathbb{C}$) in the continuous case, or $\mathcal{A} = \mathbb{F}^{N^r}$ in the discrete case. Let $\mathcal{D} = \mathbb{F}[s_1, \ldots, s_r]$ and $R \in \mathcal{D}^{g \times q}$.

Now consider the behavior

$$\mathcal{B} = \{w \in \mathcal{A}^q, \, Rw = 0\},$$

the action of R on w being given by partial derivation or shifts, respectively. A special feature of the one-dimensional (1D) case is that R can be chosen to be a full row rank matrix (essentially, this is due to the fact that if $r = 1$, then $\mathcal{D}$ is a principal ideal domain). This is not true when $r > 1$, and representations with rank deficiencies have to be considered. Now let

$$R = \begin{bmatrix} N & -D \end{bmatrix} \quad \text{and} \quad w = \begin{bmatrix} u \\ y \end{bmatrix}$$

be partitioned such that D has $p := \operatorname{rank}(R)$ columns, N has $m = q - p$ columns, and $\operatorname{rank}(R) = \operatorname{rank}(D) = p$. Note that such a partition can always be obtained if $p < q$, by a suitable permutation of the columns of R and correspondingly, of the components of w. Then the columns of N are rational combinations of the columns of D, hence

$$DH = N$$

for some rational matrix H. Thus every left annihilator of D also annihilates N. In view of the fundamental principle, this implies that u is free in the sense that for all $u \in \mathcal{A}^m$, there exists $y \in \mathcal{A}^p$ such that $Dy = Nu$; moreover, none of the components of y is free due to the rank condition on D (compare with Lemma 1.3.2). In this situation, we call

$$\mathcal{B} = \left\{ \begin{bmatrix} u \\ y \end{bmatrix} \in \mathcal{A}^{m+p}, \, Dy = Nu \right\}$$

a behavior with input-output structure, and it can be interpreted as the set of admissible input-output pairs corresponding to (D, N), where $DH = N$ and D has full column rank.

Now let a rational $p \times m$ matrix of r variables,

$$H = H(s_1, \dots, s_r) \in \mathbb{F}(s_1, \dots, s_r)^{p \times m}$$

be given. We will consider the problem of constructing a co-prime factorization of H, and our notion of a factorization is motivated by the considerations described above.

Definition 13 A pair (D, N) of polynomial matrices with the same number of rows is said to be a **left factorization** of H iff

1. $DH = N$, and
2. D has full column rank.

Similarly, a pair $(\bar{D}, \bar{N})$ of polynomial matrices with the same number of columns is called a **right factorization** of H iff $(\bar{D}^T, \bar{N}^T)$ is a left factorization of H^T, where $(\cdot)^T$ denotes transposition.

Note that we do not restrict to square matrices D. For square matrices, we recover the familiar factorizations

$$H = D^{-1}N \quad \text{and} \quad H = \bar{N}\bar{D}^{-1}.$$

Definition 14 Let (D, N) be a left factorization of H. The system of equations

$$Dy = Nu \tag{2.1}$$

is called an **input-output (IO) realization** of H. The pair $(u, y) \in \mathcal{A}^m \times \mathcal{A}^p$ is said to be an **input-output (IO) pair** with respect to (D, N) iff (2.1) is satisfied.

For a right factorization $(\bar{D}, \bar{N})$ of H,

$$u = \bar{D}v, \quad y = \bar{N}v \tag{2.2}$$

is called a **driving-variable (DV) realization** of H. If there exists a $v \in \mathcal{A}^n$ such that (2.2) is satisfied, then (u, y) is said to be an IO pair with respect to $(\bar{D}, \bar{N})$.

Equation (2.1) corresponds to a kernel representation of the set of input-output pairs with respect to (D, N):

$$(u, y) \text{ IO pair} \quad \Leftrightarrow \quad [\, N \quad -D \,] \begin{bmatrix} u \\ y \end{bmatrix} = 0.$$

Equation (2.2) yields an image representation of the set of IO pairs with respect to $(\bar{D}, \bar{N})$:

$$(u,y) \text{ IO pair} \quad \Leftrightarrow \quad \exists v: \begin{bmatrix} u \\ y \end{bmatrix} = \begin{bmatrix} \bar{D} \\ \bar{N} \end{bmatrix} v.$$

It will turn out later on that all right factorizations of H generate the same set of IO pairs. This is not true for left factorizations, however. Next, we introduce the co-primeness concept that is suitable for factorizations according to Definition 13.

2.2 Characterizations of Primeness

Definition 15 Two polynomial matrices D, N with the same number of rows are said to be **left co-prime** if the block matrix $R = \begin{bmatrix} N & -D \end{bmatrix}$ is left prime in the sense of Definition 3 (GFLP). A **left co-prime factorization** of a rational matrix H is a left factorization (D, N) of H with left co-prime matrices D, N. **Right co-primeness** and **right co-prime factorizations** are defined by transposition.

The following is implicit in the previous discussions of generalized factor primeness. We explicitly restate the observation as it will become crucial later on.

Theorem 8 *[83, 90] Two polynomial matrices D and N are left co-prime in the sense of Definition 15 iff the block matrix $R = \begin{bmatrix} N & -D \end{bmatrix}$ is a minimal left annihilator, that is, iff there exist polynomial matrices $\bar{D}$ and $\bar{N}$ such that*

1. *$N\bar{D} - D\bar{N} = 0$, and*
2. *if $N_1\bar{D} - D_1\bar{N} = 0$, we must have $D_1 = XD$ and $N_1 = XN$ for some polynomial matrix X.*

In other words, R is a minimal left annihilator of

$$M = \begin{bmatrix} \bar{D} \\ \bar{N} \end{bmatrix}.$$

Proof. We show that R is left prime iff it is a minimal left annihilator.

"if": Let R be a minimal left annihilator, say of M, and let $R = XR_1$ with $\text{rank}(R) = \text{rank}(R_1)$. Then there exists a *rational* matrix $\tilde{Y}$ such that $R_1 = \tilde{Y}R$, and hence $R_1 M = \tilde{Y}RM = 0$. But this implies $R_1 = YR$ for some polynomial matrix Y.

"only if": Let R be left prime and let M be a matrix with $RM = 0$ and $\text{rank}(R) + \text{rank}(M) = q$, where q denotes the number of columns of R. Let R^c be a minimal left annihilator of M. In particular, this implies that $\text{rank}(R^c) + \text{rank}(M) = q$. Moreover, we have $\text{im}(R) \subseteq \ker(M) = \text{im}(R^c)$, where image and kernel are defined for the operations from the left, e.g., M is interpreted as a mapping $\eta \mapsto \eta M$. This implies that there exists a polynomial matrix X such that $R = X R^c$. Note that $\text{rank}(R) = \text{rank}(R^c)$, hence the left primeness of R implies that $R^c = Y R$ for some polynomial matrix Y. But then $\text{im}(R) = \text{im}(R^c) = \ker(M)$, and thus R is a minimal left annihilator of M. $\qquad\square$

In general (D and N not necessarily left co-prime), the row module of R^c contains the row module of R, that is, $R = X R^c$ for some polynomial matrix X. Partitioning $R^c = \begin{bmatrix} N^c & -D^c \end{bmatrix}$, we have

$$D = X D^c \quad \text{and} \quad N = X N^c$$

with D^c, N^c left co-prime. The matrix X is a **maximal common left factor** of D and N in this sense, but note that X is non-unique (even if R^c is fixed). This is due to possible rank deficiencies of R^c. Again, we think of the case $r > 2$. In the 2D case, it is justified to speak of greatest common divisors, as these are unique up to unimodularity.

2.3 Constructing Co-prime Factorizations

The preceding sections provide all the necessary prerequisites for constructing a co-prime factorization of a given rational matrix $H \in \mathcal{K}^{p \times m}$, where $\mathcal{K} = \mathbb{F}(s_1, \ldots, s_r)$. Here, we restrict to left co-prime factorizations, without loss of generality. We start with the following important observation.

Theorem 9 *Let $(\bar{D}, \bar{N})$ be an arbitrary right factorization of H. Then (D, N) is a left co-prime factorization of H iff $R = \begin{bmatrix} N & -D \end{bmatrix}$ is a minimal left annihilator of*

$$M = \begin{bmatrix} \bar{D} \\ \bar{N} \end{bmatrix}.$$

Proof. "if": Left co-primeness of (D, N) is a direct consequence of Theorem 8. It remains to be shown that (D, N) is indeed a left factorization of H. As

$$N\bar{D} - D\bar{N} = 0,$$

and $H\bar{D} = \bar{N}$, we have $(N - DH)\bar{D} = 0$. As $\bar{D}$ has full row rank, $DH = N$ follows. To see that D has full column rank, note that M has $m + p$ rows and

$$\text{rank}(M) = \text{rank} \begin{bmatrix} \bar{D} \\ \bar{N} \end{bmatrix} = \text{rank} \begin{bmatrix} \bar{D} \\ H\bar{D} \end{bmatrix} = \text{rank}(\bar{D}) = m.$$

This implies that the rank of its minimal left annihilator equals p. Thus

$$p = \text{rank} \begin{bmatrix} N & -D \end{bmatrix} = \text{rank} \begin{bmatrix} DH & -D \end{bmatrix} = \text{rank}(D).$$

"only if": Let (D, N) be a left co-prime factorization of H, then

$$\left[\begin{array}{cc} N & -D \end{array}\right] M = 0.$$

As rank $\left[\begin{array}{cc} N & -D \end{array}\right] + \text{rank}\,(M) = p + m$, the matrix $\left[\begin{array}{cc} N & -D \end{array}\right]$ is already a minimal left annihilator of M according to Lemma 2.3.1 below.   □

Lemma 2.3.1. *Let $R \in \mathcal{D}^{g \times q}$ be a minimal left annihilator. If $RM = 0$ and*

$$\text{rank}\,(R) + \text{rank}\,(M) = q,$$

then R is a minimal left annihilator of M.

Proof. Let R be a minimal left annihilator of M_1. Let $\eta M_1 = 0$ for some rational row vector η. There exists a polynomial $d \neq 0$ such that $d\eta$ is polynomial, and $d\eta M_1 = 0$. Thus $d\eta = \xi R$ for some polynomial row vector ξ and hence $d\eta M = 0$ and $\eta M = 0$. Thus $\eta M_1 = 0$ implies $\eta M = 0$. As rank $(M) = \text{rank}\,(M_1)$, we conclude that

$$\eta M = 0 \quad \Leftrightarrow \quad \eta M_1 = 0 \qquad \text{for } \eta \in \mathcal{K}^{1 \times q} \supset \mathcal{D}^{1 \times q}.$$

Thus the left kernels of M and M_1 coincide, and R is a minimal left annihilator also of M.   □

In order to obtain a left co-prime factorization of H, we proceed as follows:

Algorithm 2 *Let a rational matrix $H \in \mathcal{K}^{p \times m}$ be given.*

1. *Construct a right factorization of H. This can be achieved by taking a common multiple of all denominators appearing in H, i.e., write*

$$H = \frac{\bar{N}_0}{d} \quad \text{for some } \bar{N}_0 \in \mathcal{D}^{p \times m}, \quad 0 \neq d \in \mathcal{D}.$$

Setting $\bar{D}_0 := dI_m$, we have found a right factorization $(\bar{D}_0, \bar{N}_0)$ of H (even in the classical sense, that is, $H = \bar{N}_0 \bar{D}_0^{-1}$).
2. *Now compute a minimal left annihilator of*

$$M_0 := \left[\begin{array}{c} \bar{D}_0 \\ \bar{N}_0 \end{array}\right]$$

and partition it conformably as $\left[\begin{array}{cc} N & -D \end{array}\right]$.

The following corollary is an immediate consequence of Theorem 9.

Corollary 5 *The matrix pair (D, N) as constructed above is a left co-prime factorization of H.*

2.4 Properties of Co-prime Factorizations

First, we collect some facts on **determinantal ideals** [47, p. 6] that will be needed in the following sections. For a polynomial matrix R, and an integer $\nu \geq 0$, let

$$\mathcal{I}_\nu(R) \subseteq \mathcal{D} = \mathbb{F}[s_1, \ldots, s_r]$$

denote the polynomial ideal generated by the $\nu \times \nu$ minors of R. By convention, $\mathcal{I}_0(R) = \mathcal{D}$, and $\mathcal{I}_\nu(R) = 0$ if ν exceeds the number of rows or columns of R. Hence

$$\mathcal{I}_0(R) \supseteq \mathcal{I}_1(R) \supseteq \mathcal{I}_2(R) \supseteq \cdots$$

The largest ν such that $\mathcal{I}_\nu(R) \neq 0$ is equal to the rank of R. Define

$$\mathcal{I}(R) := \mathcal{I}_{\text{rank}(R)}(R).$$

Recall that for a polynomial ideal $\mathcal{I} \neq \mathcal{D}$, its **co-dimension** (height) coincides with its **grade** (depth) [39, p. 187]. The co-dimension is an integer between zero and r, measuring the "size" of a polynomial ideal in the sense that $\text{codim}(0) = 0$ and

$$\mathcal{I}_1 \subseteq \mathcal{I}_2 \quad \Rightarrow \quad \text{codim}(\mathcal{I}_1) \leq \text{codim}(\mathcal{I}_2).$$

In the case when $\mathbb{F} = \mathbb{R}$ or $\mathbb{F} = \mathbb{C}$, one can think of $\text{codim}(\mathcal{I})$ as the usual co-dimension in $\mathbb{C}^r$ of the associated variety

$$\mathcal{V}(\mathcal{I}) = \{\xi \in \mathbb{C}^r, \ f(\xi) = 0 \text{ for all } f \in \mathcal{I}\}.$$

We set $\text{codim}(\mathcal{D}) = \infty$ and hence we have

$$\infty = \text{codim}(\mathcal{I}_0(R)) \geq \text{codim}(\mathcal{I}_1(R)) \geq \ldots \geq \text{codim}(\mathcal{I}(R)) \geq 1$$

and $\text{codim}(\mathcal{I}_\nu(R)) = 0$ for all $\nu > \text{rank}(R)$.

Theorem 10 *Let (D, N) be a left co-prime factorization of H, and let (D_1, N_1) be an arbitrary left factorization of H. Then there exists a polynomial matrix X such that*

$$D_1 = XD \quad \text{and} \quad N_1 = XN.$$

In particular, $\mathcal{I}_\nu(D_1) \subseteq \mathcal{I}_\nu(D)$ for all $\nu \geq 0$, and $\mathcal{I}(D_1) \subseteq \mathcal{I}(D)$.

Proof. According to Theorem 9, $\begin{bmatrix} N & -D \end{bmatrix}$ is a minimal left annihilator of M_0, which is defined as in Algorithm 2. In view of Theorem 8, all we need to show is, $N_1 \bar{D}_0 - D_1 \bar{N}_0 = 0$. But this follows directly from $D_1 H = D_1 \bar{N}_0 \bar{D}_0^{-1} = N_1$. The statement on the determinantal ideals follows from [47, p. 7]

$$\mathcal{I}_\nu(D_1) = \mathcal{I}_\nu(XD) \subseteq \mathcal{I}_\nu(D) \quad \text{for all } \nu \geq 0,$$

and the fact that $\text{rank}(D) = \text{rank}(D_1)$. $\qquad\square$

Thus the determinantal ideal of D in a left co-prime factorization is maximal among the determinantal ideals of denominator matrices in left factorizations. This should be compared with the following classical result from 1D theory: If $H = D^{-1}N$ is a left co-prime factorization, and $H = D_1^{-1}N_1$ an arbitrary left factorization, then $\det(D)$ divides $\det(D_1)$. In other words, the principal ideal $\mathcal{I}(D) = \langle\det(D)\rangle$ contains $\mathcal{I}(D_1) = \langle\det(D_1)\rangle$.

Due to Theorem 10, we can characterize the relation between two left co-prime factorizations of one and the same rational matrix as follows.

Corollary 6 *Let (D,N) and (D_1,N_1) be two left co-prime factorizations of H. Then there exist polynomial matrices X and Y such that*

$$D_1 = XD \quad and \quad N_1 = XN, \qquad D = YD_1 \quad and \quad N = YN_1.$$

In other words, the rows of $\begin{bmatrix} N & -D \end{bmatrix}$ and $\begin{bmatrix} N_1 & -D_1 \end{bmatrix}$ generate the same polynomial module. In particular, $\mathcal{I}_\nu(D) = \mathcal{I}_\nu(D_1)$ for all $\nu \geq 0$ and $\mathcal{I}(D) = \mathcal{I}(D_1)$.

Next, we turn to the sets of input-output pairs associated with left and right factorizations according to Definition 14.

Lemma 2.4.1. *The set of IO pairs with respect to a left co-prime factorization of H equals the set of IO pairs with respect to any right factorization of H.*

Proof. Let (D,N) be a left co-prime factorization of H, and let $(\bar{D},\bar{N})$ be an arbitrary right factorization of H. Then $R = \begin{bmatrix} N & -D \end{bmatrix}$ is a minimal left annihilator of $M = \begin{bmatrix} \bar{D}^T & \bar{N}^T \end{bmatrix}^T$, hence

$$Dy = Nu \quad \Leftrightarrow \quad \exists v : u = \bar{D}v, \, y = \bar{N}v$$

according to the fundamental principle. □

Corollary 7 *Let (D,N) be a left co-prime factorization of H, and let (D_1,N_1) be an arbitrary left factorization of H. Then*

$$Dy = Nu \quad \Rightarrow \quad D_1y = N_1u, \qquad (2.3)$$

that is, an IO pair with respect to (D,N) is also an IO pair with respect to (D_1,N_1). Thus a left co-prime factorization generates the smallest set of IO pairs associated with any left factorization of H. Moreover, (D_1,N_1) is left co-prime iff the converse of (2.3) holds.

Proof. In view of Theorem 10, we only need to prove that if equivalence holds in (2.3), then also (D_1,N_1) is left co-prime. In view of the preceding lemma, let $(\bar{D},\bar{N})$ be an arbitrary right factorization of H, then

$$D_1y = N_1u \quad \Leftrightarrow \quad Dy = Nu \quad \Leftrightarrow \quad \exists v : u = \bar{D}v, \, y = \bar{N}v.$$

Oberst's duality theorem (*i.e.*, the converse of the fundamental principle) implies that $\begin{bmatrix} N_1 & -D_1 \end{bmatrix}$ is a minimal left annihilator, hence left co-prime. □

The co-prime factorization is illustrated by the following example.

Example 2.4.1. Consider

$$H = \begin{bmatrix} s_2/s_1 \\ s_3/s_1 \end{bmatrix} \in \mathbb{R}(s_1, s_2, s_3)^{2\times 1}, \quad \bar{N}_0 = \begin{bmatrix} s_2 \\ s_3 \end{bmatrix}, \quad \bar{D}_0 = [s_1].$$

Computing a minimal left annihilator of

$$M_0 = \begin{bmatrix} s_1 \\ s_2 \\ s_3 \end{bmatrix}$$

yields the left co-prime factorization

$$[\ N\ |\ -D\ ] = \begin{bmatrix} 0 & -s_3 & s_2 \\ s_3 & 0 & -s_1 \\ -s_2 & s_1 & 0 \end{bmatrix}.$$

We have $\mathcal{I}(D) = \langle s_1^2, s_1 s_2, s_1 s_3 \rangle$. Note that $(\bar{D}_0, \bar{N}_0)$ is already a right co-prime factorization of H and that $\mathcal{I}(\bar{D}_0) = \langle s_1 \rangle \neq \mathcal{I}(D)$.

This phenomenon was already observed by Lin [41] for classical factorizations of rational matrices in $r > 2$ indeterminates. It signifies that another classical result fails to generalize to the multidimensional setting: For 1D systems, we have $\det(D) = \det(\bar{D})$ (up to non-zero constant factors) for a right co-prime factorization $(\bar{D}, \bar{N})$ and a left co-prime factorization (D, N) of $H = D^{-1}N = \bar{N}\bar{D}^{-1}$. It is known however [46] that the result still holds for 2D systems, and we will shortly recover this fact using different methods. It will turn out that even for dimensions $r > 2$, there are some interesting connections between the determinantal ideals associated to right and left co-prime factorizations.

2.5 Factorization Ideals

We start by quoting a fundamental result from the theory of factorization ideals as developed by Buchsbaum and Eisenbud [14]; see also Northcott [47, Ch. 7].

Theorem 11 *[14, 47] Let (D, N) be a left co-prime factorization of H, and let $(\bar{D}, \bar{N})$ be an arbitrary right factorization of H. As usual, let*

$$R = [\ N\ \ -D\ ] \quad and \quad M = \begin{bmatrix} \bar{D} \\ \bar{N} \end{bmatrix}.$$

Then there exist ideals $\mathcal{J}_0, \mathcal{J}_1, \mathcal{J}_2 \subseteq \mathcal{D}$ such that

1. $\mathcal{I}(M) = \mathcal{J}_1 \mathcal{J}_0$;
2. $\mathcal{I}(R) = \mathcal{J}_2 \mathcal{J}_1$;
3. $\mathrm{codim}(\mathcal{J}_i) \geq i + 1$ for $i = 0, 1, 2$.

Furthermore, if M has n columns and rank m, then $\mathcal{J}_0$ can be generated by $\binom{n}{m}$ elements. A particularly interesting situation arises when M has full column rank (e.g., if $M = M_0$ as in Algorithm 2). Then $\mathcal{J}_0$ is a principal ideal, moreover, it is the smallest principal ideal containing $\mathcal{I}(M)$.

We only present a small ingredient of the proof of Theorem 11 that is crucial for our derived results below: Let R be a minimal left annihilator of M, then there exists a finite free resolution

$$0 \to F_l \to \ldots \to F_2 := \mathcal{D}^{1 \times g} \xrightarrow{R} F_1 := \mathcal{D}^{1 \times q} \xrightarrow{M} F_0 := \mathcal{D}^{1 \times n} \to \mathrm{cok}(M) \to 0$$

of $\mathrm{cok}(M) = \mathcal{D}^{1 \times n} / \mathcal{D}^{1 \times q} M$. Let $M^{(m)}$ denote the m-th exterior power [47, p. 4] of $M \in \mathcal{D}^{q \times n}$, that is a $\binom{q}{m} \times \binom{n}{m}$ matrix whose entries are the $m \times m$ minors of M. Each row of $M^{(m)}$ corresponds to a particular choice of m rows of M, say those with row indices

$$1 \leq i_1 < \ldots < i_m \leq q. \tag{2.4}$$

For the sake of uniqueness, we assume that these m-tuples are ordered lexicographically; and analogously for the columns of $M^{(m)}$. As $\mathrm{rank}(M) = m$,

$$\mathcal{I}(M) = \mathcal{I}_m(M) = \mathcal{I}_1(M^{(m)}).$$

Similarly for $R \in \mathcal{D}^{g \times q}$ with $\mathrm{rank}(R) = p = q - m$, we have $\mathcal{I}(R) = \mathcal{I}_p(R) = \mathcal{I}_1(R^{(p)})$, and $R^{(p)}$ has $\binom{g}{p}$ rows and $\binom{q}{p} = \binom{q}{m}$ columns. Now, in the situation of Theorem 11, there exist column vectors U and W, and row vectors V, Z such that

$$R^{(p)} = UV \quad \text{and} \quad M^{(m)} = WZ. \tag{2.5}$$

Moreover, the entries of W and V are the same (up to non-zero constant factors), indeed the entry of W that corresponds to the rows of M with row indices (2.4) equals (up to a non-zero constant factor) the entry of V that corresponds to the p columns of R with column indices

$$\{1, \ldots, q\} \setminus \{i_1, \ldots, i_m\}.$$

In particular, $\mathcal{I}_1(W) = \mathcal{I}_1(V) =: \mathcal{J}_1$. With $\mathcal{J}_0 := \mathcal{I}_1(Z)$, $\mathcal{J}_2 := \mathcal{I}_1(U)$, we have

$$\mathcal{I}(M) = \mathcal{I}_1(WZ) = \mathcal{I}_1(W)\mathcal{I}_1(Z) = \mathcal{J}_1 \mathcal{J}_0,$$

and similarly, $\mathcal{I}(R) = \mathcal{J}_2 \mathcal{J}_1$.

Example 2.5.1. Consider once more

$$R = \begin{bmatrix} 0 & -s_3 & s_2 \\ s_3 & 0 & -s_1 \\ -s_2 & s_1 & 0 \end{bmatrix} \quad \text{and} \quad M = \begin{bmatrix} s_1 \\ s_2 \\ s_3 \end{bmatrix}.$$

Factorizations according to (2.5) are given by

$$R^{(2)} = \begin{bmatrix} s_3^2 & -s_2 s_3 & s_1 s_3 \\ -s_2 s_3 & s_2^2 & -s_1 s_2 \\ s_1 s_3 & -s_1 s_2 & s_1^2 \end{bmatrix} = \begin{bmatrix} s_3 \\ -s_2 \\ s_1 \end{bmatrix} \begin{bmatrix} s_3 & -s_2 & s_1 \end{bmatrix}$$

and $M^{(1)} = M$ has the trivial factorization $W = M$, $Z = 1$. Then $\mathcal{J}_0 = \mathcal{D} = \mathbb{R}[s_1, s_2, s_3]$ and $\mathcal{J}_1 = \mathcal{J}_2 = \langle s_1, s_2, s_3 \rangle$. Alternatively, one could find these ideals by looking at $\mathcal{I}(R) = \langle s_1^2, s_1 s_2, s_1 s_3, s_2^2, s_2 s_3, s_3^2 \rangle$ and $\mathcal{I}(M) = \langle s_1, s_2, s_3 \rangle$, and using the characterizations given in Theorem 11. The smallest principal ideal that contains the zero-dimensional ideal $\mathcal{I}(M)$ is the whole of $\mathcal{D}$. Thus $\mathcal{J}_0 = \langle 1 \rangle = \mathcal{D}$, and $\mathcal{J}_1 = \mathcal{J}_2 = \mathcal{I}(M)$.

Corollary 8 *Let* (D, N), $(\bar{D}, \bar{N})$, R, M, *and* $\mathcal{J}_0$, $\mathcal{J}_1$, $\mathcal{J}_2$ *be as in Theorem 11. Then there exists a principal ideal* $\mathcal{J}_1' \subseteq \mathcal{J}_1$ *such that*

1. $\mathcal{I}(\bar{D}) = \mathcal{J}_1' \mathcal{J}_0$;
2. $\mathcal{I}(D) = \mathcal{J}_2 \mathcal{J}_1'$.

Moreover, $\mathcal{J}_1'$ *is generated by the greatest common divisor of the* $p \times p$ *minors of* D, *that is,* $\mathcal{J}_1'$ *is the smallest principal ideal containing* $\mathcal{I}(D)$.

Proof. The p-th exterior power of D is the last column of $R^{(p)}$. Similarly, $\bar{D}^{(m)}$ is the first row of $M^{(m)}$. In view of (2.5),

$$\bar{D}^{(m)} = w_1 Z,$$

where w_1 denotes the first component of W. Up to a non-zero constant factor, w_1 is also the last component of V, hence

$$D^{(p)} = U w_1 \qquad (2.6)$$

(up to non-zero constant factors). Hence with $\mathcal{J}_1' := \langle w_1 \rangle \subseteq \mathcal{I}_1(W) = \mathcal{J}_1$, we obtain

$$\mathcal{I}(\bar{D}) = \mathcal{I}_m(\bar{D}) = \mathcal{I}_1(\bar{D}^{(m)}) = \langle w_1 \rangle \mathcal{I}_1(Z) = \mathcal{J}_1' \mathcal{J}_0.$$

Similarly, $\mathcal{I}(D) = \mathcal{J}_2 \mathcal{J}_1'$. Finally from (2.6), w_1 is certainly a common divisor of the entries of $D^{(p)}$. Suppose that the entries of U possess a non-constant common factor, say h. Then $\mathcal{I}_1(U) \subseteq \langle h \rangle$, and hence $\text{codim}(\mathcal{I}_1(U)) \leq \text{codim}(\langle h \rangle) = 1$. But we know that $\text{codim}(\mathcal{J}_2) = \text{codim}(\mathcal{I}_1(U)) \geq 3$, and hence we conclude that the entries of U are devoid of a non-constant common factor. Thus w_1 is the greatest common divisor of the $p \times p$ minors of D. $\qquad \square$

Example 2.5.2. In Example 2.4.1, we have $\mathcal{I}(D) = \langle s_1^2, s_1 s_2, s_1 s_3 \rangle$, hence $\mathcal{J}_1' = \langle s_1 \rangle$. Note that $\mathrm{codim}(\mathcal{J}_1') = 1$ (compare this with Statement 3 in Theorem 11).

It is worth noting that in the situation of Theorem 11 and Corollary 8, the ideals $\mathcal{I}(R)$, $\mathcal{I}(D)$, and thus also $\mathcal{J}_1'$ are uniquely determined by H.

2.6 The Bivariate Case

Hilbert's syzygy theorem [39, p. 208] implies that for $\mathcal{D} = \mathbb{F}[s_1, s_2]$, $\mathbb{F}$ a field, any finitely generated $\mathcal{D}$-module $\mathcal{M}$ possesses a finite free resolution of length $l \leq 2$. The following fact is implicit in [48, p. 148]. As it is crucial for the considerations below, we shortly restate it.

Theorem 12 *[48] A minimal left (right) annihilator of a bivariate polynomial matrix is either zero or can be chosen to be a matrix with full row (column) rank.*

Proof. Let $M \in \mathcal{D}^{q \times n}$ be given. Let

$$\ker(M) = \{\eta \in \mathcal{D}^{1 \times q}, \eta M = 0\} \quad \text{and} \quad \mathrm{cok}(M) = \mathcal{D}^{1 \times n}/\mathcal{D}^{1 \times q} M.$$

A polynomial matrix is a minimal left annihilator of M iff its rows generate $\ker(M)$. The sequence

$$0 \to \ker(M) \hookrightarrow \mathcal{D}^{1 \times q} \xrightarrow{M} \mathcal{D}^{1 \times n} \to \mathrm{cok}(M) \to 0$$

is exact. As $\mathrm{cok}(M)$ is a finitely generated $\mathcal{D}$-module, there exists a finite free resolution of $\mathrm{cok}(M)$ of length ≤ 2. We conclude that $\ker(M)$ is free, which yields the desired result: Either $\ker(M) = 0$, which corresponds to a full row rank matrix M, whose left annihilators are necessarily zero, or $\ker(M) \neq 0$, in which case it can be spanned by a set of linearly independent row vectors. The assertion on minimal right annihilators follows by transposition. $\square$

It is important to note that in spite of Theorem 12, it does not suffice to solve the linear system of equations $\eta M = 0$ over the field $\mathcal{K} = \mathbb{F}(s_1, s_2)$ when searching a minimal left annihilator of M. This is true although we can choose every generator η of $\ker(M)$ such that the polynomials $\eta_1, \ldots, \eta_q$ are co-prime.

Example 2.6.1. The left kernel of

$$M = \begin{bmatrix} s_1 + 1 \\ s_2 \\ s_1 s_2 \end{bmatrix}$$

is generated (over $\mathcal{K}$) by

$$\tilde{R} = \begin{bmatrix} -s_2 & s_1 + 1 & 0 \\ -s_1 s_2 & 0 & s_1 + 1 \end{bmatrix}.$$

A minimal left annihilator of M is given by

$$R = \begin{bmatrix} -s_2 & s_1 + 1 & 0 \\ -s_2 & 1 & 1 \end{bmatrix}.$$

Its row module is strictly greater than that of $\tilde{R}$. This is of course due to the fact that R is minor left prime, whereas $\tilde{R}$ is not.

Recall that any polynomial matrix R can be factored as $R = X R^c$ with R^c left prime. In view of Theorem 12, the matrix R^c can be assumed to have full row rank in the bivariate situation. Then X is unique up to post-multiplication by a unimodular matrix. Thus it is justified to speak of the **greatest left factor** of a bivariate matrix. Similarly, two bivariate polynomial matrices D and N with the same number of rows possess a well-defined **greatest common left factor** (GCLF).

Another consequence of Theorem 12 is that the minimal left annihilator of M_0 in Algorithm 2 can be chosen to be a full row rank matrix. Then D is a square matrix, and the left factorization is $H = D^{-1}N$. This is the fundamental reason why it is justified to restrict to classical factorizations (with square denominator matrix), in the 2D just like in the 1D case. Many authors, including Morf *et al.* [46], have used classical matrix fraction descriptions for 2D transfer functions. This section aims at pointing out why this approach works nicely for 2D systems, but is bound to become inadequate in dimensions greater than two.

The co-prime factorization is illustrated by the following example.

Example 2.6.2. Consider

$$H = \frac{1}{s_1 - s_2 - 1} \begin{bmatrix} s_1 - s_1 s_2 - 1 + s_2 - s_2^2 & -2s_2 \\ -s_1 + s_1 s_2 + s_2^2 + s_2 & s_1 + s_2 + 1 \end{bmatrix}$$

and let

$$\bar{N}_0 = \begin{bmatrix} s_1 - s_1 s_2 - 1 + s_2 - s_2^2 & -2s_2 \\ -s_1 + s_1 s_2 + s_2^2 + s_2 & s_1 + s_2 + 1 \end{bmatrix}, \quad \bar{D}_0 = (s_1 - s_2 - 1)I_2.$$

Computing a minimal left annihilator of

$$M_0 = \begin{bmatrix} \bar{D}_0 \\ \bar{N}_0 \end{bmatrix}$$

yields the left co-prime factorization

$$R = [\, N \mid -D \,] = \begin{bmatrix} -s_2 + 1 & s_2 & -s_2 - 1 & -s_2 \\ s_2 & s_1 + 1 & -s_1 & -s_1 + 1 \end{bmatrix}.$$

As the 2×2 minors of R generate the whole of $\mathcal{D} = \mathbb{R}[s_1, s_2]$, we have $\mathcal{I}(R) = \mathcal{D}$. Note that

$$\mathcal{I}(D) = \langle s_1 - s_2 - 1 \rangle.$$

Furthermore $(\bar{D}_0, \bar{N}_0)$ is not right co-prime and $\mathcal{I}(\bar{D}_0) = \langle (s_1 - s_2 - 1)^2 \rangle \neq \mathcal{I}(D)$. The relation between these determinantal ideals will be discussed below.

This example is closely related to the one given in [4]. In fact, $H = A^{-1}B$, where

$$A = \begin{bmatrix} s_1^2 + 2s_1 s_2 + s_1 & s_1^2 - s_1 + 2s_1 s_2 - s_2 \\ 2s_1 s_2 + s_1 + s_2 + 1 & 2s_1 s_2 \end{bmatrix}$$

and

$$B = \begin{bmatrix} s_1 + s_2^2 & s_1^2 + s_1 + 2s_1 s_2 + s_2 \\ s_2^2 - s_1 s_2 + s_1 - s_2 + 1 & 2s_2(s_1 + 1) \end{bmatrix}.$$

Thus N and D can be seen as factors of A and B after extraction of their greatest common left factor, in fact

$$A = \begin{bmatrix} s_1 & s_1 + s_2 \\ s_1 + 1 & s_2 \end{bmatrix} D \quad \text{and} \quad B = \begin{bmatrix} s_1 & s_1 + s_2 \\ s_1 + 1 & s_2 \end{bmatrix} N.$$

Comparing with the result given in [4], we find that the GCLF given there, namely

$$\begin{bmatrix} s_2 & s_1 \\ s_2 - s_1 - 1 & s_1 + 1 \end{bmatrix} = \begin{bmatrix} s_1 & s_1 + s_2 \\ s_1 + 1 & s_2 \end{bmatrix} \begin{bmatrix} -1 & 1 \\ 1 & 0 \end{bmatrix}$$

differs from the one computed above only by a post-multiplication by a unimodular matrix.

The characterizations of co-prime factorizations in terms of determinantal ideals that were derived in the preceding sections have nice special versions in the bivariate situation.

The co-dimension of an ideal $\mathcal{I} \subset \mathcal{D} = \mathbb{F}[s_1, s_2]$ is either zero (corresponds to $\mathcal{I} = 0$), or one, or two (corresponds to a zero-dimensional $\mathcal{I}$, i.e., one for which $\mathcal{D}/\mathcal{I}$ is finite-dimensional as an $\mathbb{F}$-vector space). The co-dimension of $\mathcal{D}$ itself is set to be ∞ as usual. Obviously, $\text{codim}(\mathcal{I}) \geq 3$ already implies that $\mathcal{I} = \mathcal{D}$.

If R has full row rank, the condition $\text{codim}(\mathcal{I}(R)) \geq 2$ is equivalent to minor (and thus classical factor [46]) left primeness of R. As mentioned in Chapter 1, Wood et al. [83] showed that then the claim of Theorem 2 is actually an equivalence. This implies that generalized factor left primeness coincides with minor and classical factor left primeness in the case of full row rank bivariate matrices.

Corollary 9 *[14, 47] Let (D, N) be a left co-prime factorization of H, and let $(\bar{D}, \bar{N})$ be an arbitrary right factorization of H. As usual, let*

$$R = [\, N \quad -D \,] \quad and \quad M = \begin{bmatrix} \bar{D} \\ \bar{N} \end{bmatrix}.$$

Then there exists an ideal $\mathcal{J}_0 \subseteq \mathcal{D}$ such that

$$\mathcal{I}(M) = \mathcal{I}(R)\mathcal{J}_0.$$

Furthermore, if M has n columns and rank m, then $\mathcal{J}_0$ can be generated by $\binom{n}{m}$ elements. A particularly interesting situation arises when M has full column rank (e.g., if $M = M_0$ as in Algorithm 2). Then $\mathcal{J}_0$ is a principal ideal, moreover, it is the smallest principal ideal containing $\mathcal{I}(M)$.

Example 2.6.3. Consider once more

$$R = \begin{bmatrix} -s_2 + 1 & s_2 & -s_2 - 1 & -s_2 \\ s_2 & s_1 + 1 & -s_1 & -s_1 + 1 \end{bmatrix}$$

and

$$M_0 = \begin{bmatrix} s_1 - s_2 - 1 & 0 \\ 0 & s_1 - s_2 - 1 \\ s_1 - s_1 s_2 - 1 + s_2 - s_2^2 & -2 s_2 \\ -s_1 + s_1 s_2 + s_2^2 + s_2 & s_1 + s_2 + 1 \end{bmatrix}$$

from the previous example. Factorizations according to (2.5) are given by

$$M_0^{(2)} = \begin{bmatrix} s_1 - s_2 - 1 \\ -2 s_2 \\ s_1 + s_2 + 1 \\ s_2^2 + s_1 s_2 - s_2 - s_1 + 1 \\ s_1 - s_2 - s_1 s_2 - s_2^2 \\ s_1 + 1 - s_2 - s_1 s_2 - s_2^2 \end{bmatrix} (s_1 - s_2 - 1)$$

and

$$R^{(2)} = \big[s_1 + 1 - s_2 - s_1 s_2 - s_2^2\,, \; -(s_1 - s_2 - s_1 s_2 - s_2^2),$$
$$s_2^2 + s_1 s_2 - s_2 - s_1 + 1\,, \; s_1 + s_2 + 1\,, \; 2 s_2\,, \; s_1 - s_2 - 1\big]$$

which has the trivial factorization $U = 1$, $V = R^{(2)}$. Recall that $\mathcal{I}(R) = \mathcal{D}$. Now $\mathcal{I}(M_0) = \mathcal{J}_0 = \langle s_1 - s_2 - 1 \rangle$.

Corollary 10 *Let (D, N), $(\bar{D}, \bar{N})$, R, M, and $\mathcal{J}_0$ be as in Corollary 9. Then $\mathcal{I}(D)$ is a principal ideal and*

$$\mathcal{I}(\bar{D}) = \mathcal{I}(D)\mathcal{J}_0.$$

Proof. The p-th exterior power of D is the last column of $R^{(p)}$ (up to the sign). Similarly, $\bar{D}^{(m)}$ is the first row of $M^{(m)}$. In view of (2.5),

$$\bar{D}^{(m)} = w_1 Z,$$

where w_1 denotes the first component of W. Up to a constant factor, w_1 is also the last component of V, hence

$$D^{(p)} = U w_1$$

(up to constant factors). On the other hand, as $\mathrm{codim}(\mathcal{J}_2) = \mathrm{codim}(\mathcal{I}_1(U)) \geq 3$, the entries of U generate the whole of $\mathcal{D}$, thus $\mathcal{I}(D) = \mathcal{I}_p(D) = \mathcal{I}_1(D^{(p)}) = \langle w_1 \rangle$. Hence

$$\mathcal{I}(\bar{D}) = \mathcal{I}_m(\bar{D}) = \mathcal{I}_1(\bar{D}^{(m)}) = \langle w_1 \rangle \mathcal{I}_1(Z) = \mathcal{I}(D)\mathcal{J}_0.$$

It remains to be noted that $\mathcal{I}(D)$ is generated by w_1, which is the greatest common divisor of the $p \times p$ minors of D.   □

Example 2.6.4. In the example discussed above, we have $\mathcal{I}(D) = \mathcal{J}_0 = \langle s_1 - s_2 - 1 \rangle$, and accordingly, $\mathcal{I}(\bar{D}) = \langle (s_1 - s_2 - 1)^2 \rangle$.

Finally we turn to doubly co-prime factorizations and recover a result that is well-known in the 1D setting, and has been proven to hold for classical factorizations in the 2D case [46].

Corollary 11 *Let* (D, N) *be a left co-prime factorization of* H, *and let* $(\bar{D}, \bar{N})$ *be a right co-prime factorization of* H. *Let* R *and* M *be defined as usual. Then*

$$\mathcal{I}(R) = \mathcal{I}(M) \quad and \quad \mathcal{I}(D) = \mathcal{I}(\bar{D}).$$

Proof. Double co-primeness means: R is a minimal left annihilator of M, and M is a minimal right annihilator of R. Thus the sequences

$$\mathcal{D}^{1 \times g} \xrightarrow{R} \mathcal{D}^{1 \times q} \xrightarrow{M} \mathcal{D}^{1 \times n} \quad and \quad \mathcal{D}^{1 \times n} \xrightarrow{M^T} \mathcal{D}^{1 \times q} \xrightarrow{R^T} \mathcal{D}^{1 \times g}$$

are exact and they can be extended as finite free resolutions of $\mathrm{cok}(M)$ and $\mathrm{cok}(R^T)$, respectively. Consider the factorizations according to (2.5), let $R^{(p)} = UV$, $M^{(m)} = WZ$ and, $(M^T)^{(m)} = \tilde{U}\tilde{V}$, $(R^T)^{(p)} = \tilde{W}\tilde{Z}$ with

$$\mathcal{I}_1(W) = \mathcal{I}_1(V) \quad and \quad \mathcal{I}_1(\tilde{W}) = \mathcal{I}_1(\tilde{V}).$$

Recall that we must have $\mathcal{I}_1(U) = \mathcal{I}_1(\tilde{U}) = \mathcal{D}$. Now taking into account that $(R^T)^{(p)} = (R^{(p)})^T$,

$$\mathcal{I}(R) = \mathcal{I}_1(U)\mathcal{I}_1(V) = \mathcal{I}_1(V) = \mathcal{I}_1(W) = \mathcal{I}_1(\tilde{W})\mathcal{I}_1(\tilde{Z})$$

and

$$\mathcal{I}(M) = \mathcal{I}_1(\tilde{U})\mathcal{I}_1(\tilde{V}) = \mathcal{I}_1(\tilde{V}) = \mathcal{I}_1(\tilde{W}) = \mathcal{I}_1(W)\mathcal{I}_1(Z),$$

hence $\mathcal{I}(R) = \mathcal{I}(M)\mathcal{I}_1(\tilde{Z})$ and $\mathcal{I}(M) = \mathcal{I}(R)\mathcal{I}_1(Z)$. We conclude that $\mathcal{I}(R) = \mathcal{I}(M)$ and the statement for the denominator matrices D and $\bar{D}$ follows.   □

3. Stabilizability

A variety of design problems can be solved using the well-known parameterization of all stabilizing controllers of a plant. This parameterization, first introduced by Youla *et al.* [87] in 1976, is now a fundamental ingredient of almost every textbook on robust or optimal control. It is based on co-prime factorizations of the plant, which is represented by a rational transfer matrix, over the ring of stable rational functions.

For one-dimensional (1D) systems, *i.e.*, univariate rational functions, co-primeness admits a characterization in terms of Bézout relations, and these are essential for deriving the desired parameterization. In the presence of more than one independent variable, there are several notions of co-primeness: Zero co-primeness is a strong form which admits a "1D-like" Bézout relation. Minor co-primeness is a weaker form which gives rise to a generalized type of Bézout relation. However, the concept is still restrictive as there are rational matrices that do not possess a minor co-prime factorization (an example is given below). Therefore, we will use the alternative notion of a co-prime factorization developed in the previous chapter. It guarantees the existence of co-prime factorizations for arbitrary rational matrices.

Within this framework, we give necessary and sufficient conditions for stabilizability. Analogously to the 1D case, the existence of a zero co-prime factorization of the plant is sufficient for stabilizability. This has been noted by various authors [82, 86]. In the one- and two-dimensional case, the condition is also necessary. For 2D systems, this result was established by Guiver and Bose [30]. For rational matrices of more than two variables, however, we shall only be able to show that stabilizability is a property that lies "between" freeness and projectivity of certain modules associated to the plant. This result is due to Sule [73] who obtains it within the very comprehensive setting of systems over commutative rings. Using different methods, Lin [41] derived another necessary and sufficient condition for stabilizability of n-dimensional systems.

3.1 Co-primeness over Rings of Fractions

Let $\mathbb{F}$ be a field, and let $\mathcal{P} := \mathbb{F}[z]$ denote the polynomial ring in the indeterminates $z = (z_1, \ldots, z_n)$. Let $\mathcal{Q}$ denote a multiplicatively closed subset of

$\mathcal{P}$, in particular, $1 \in \mathcal{Q}$. For simplicity, we assume $0 \notin \mathcal{Q}$. Define the ring of fractions of $\mathcal{P}$ with denominator set $\mathcal{Q}$ by

$$\mathcal{P}_\mathcal{Q} = \left\{ \frac{p}{q} \mid p \in \mathcal{P}, q \in \mathcal{Q} \right\}.$$

By means of the canonical injection $\mathcal{P} \to \mathcal{P}_\mathcal{Q}$, $p \mapsto \frac{p}{1}$, we identify $\mathcal{P}$ as a sub-ring of $\mathcal{P}_\mathcal{Q}$. Naturally, $\mathcal{P}_\mathcal{Q}$ is in turn a sub-ring of $\mathcal{R} := \mathbb{F}(z)$, the field of rational functions of z.

Example 3.1.1.

1. Let $\mathcal{Q}$ denote the set of polynomials with a non-zero constant term. Then $\mathcal{C} := \mathcal{P}_\mathcal{Q}$ is the ring of causal rational functions.
2. Suppose that $\mathbb{F} = \mathbb{R}$ or $\mathbb{F} = \mathbb{C}$. Take $\mathcal{Q}$ to be

$$\mathcal{Q} = \{p \in \mathcal{P} \mid p(\zeta) \neq 0 \ \forall \zeta \in \bar{U}^n\},$$

where

$$\bar{U} = \{\zeta \in \mathbb{C} \mid |\zeta| \leq 1\}.$$

Then $\mathcal{P}_\mathcal{Q}$ is the set of structurally (or internally) stable [30] functions. A weaker stability concept is obtained when

$$\mathcal{Q} = \{p \in \mathcal{P} \mid p(0, \dots, 0, \zeta_j, 0, \dots, 0) \neq 0 \ \forall 1 \leq j \leq n, \ \zeta_j \in \bar{U}\}.$$

The elements of $\mathcal{P}_\mathcal{Q}$ are called practically BIBO (bounded input, bounded output) stable [1].

In the following, let $\mathcal{S}$ denote a ring of fractions of $\mathcal{P}$, with denominator set $\mathcal{Q}$. Clearly, the set of units of $\mathcal{S}$ is

$$U(\mathcal{S}) = \left\{ \frac{p}{q} \mid p, q \in \mathcal{Q} \right\}.$$

Note that $\mathcal{S}$ is a factorial ring [39, p. 81], in fact, the irreducible factors of $s = \frac{p}{q} \in \mathcal{S}$ are precisely the irreducible factors of p over $\mathcal{P}$. This is due to the fact that s and p are associated, that is, they differ only by a factor in $U(\mathcal{S})$. Thus any two elements of $\mathcal{S}$ possess a well-defined greatest common divisor (gcd). Two elements of $\mathcal{S}$ are said to be *co-prime* if their gcd is in $U(\mathcal{S})$. The greatest common divisor of two polynomials over $\mathcal{P}$ is denoted by $\gcd_\mathcal{P}$. The following lemma summarizes these considerations.

Lemma 3.1.1. *For $i = 1, 2$, let $s_i = \frac{p_i}{q_i} \in \mathcal{S}$. Then*

$$\gcd(s_1, s_2) = \gcd_\mathcal{P}(p_1, p_2).$$

In particular, s_1 and s_2 are co-prime iff $\gcd_\mathcal{P}(p_1, p_2) \in \mathcal{Q}$.

As usual, the concept of gcd can easily be extended to any finite number of elements of S, and we write $\gcd(s_1, \ldots, s_k)$ for $k \geq 1$. Co-primeness can also be characterized in terms of generalized Bézout relations.

Lemma 3.1.2. *Two elements $s_1, s_2 \in S$ are co-prime iff for all $1 \leq j \leq n$, there exist $t_1^{(j)}, t_2^{(j)} \in S$ such that*

$$t_1^{(j)} s_1 + t_2^{(j)} s_2 = p^{(j)} \tag{3.1}$$

where $0 \neq p^{(j)} \in \mathcal{P}$ is independent of z_j.

We write $z^{(j)} := (z_1, \ldots, z_{j-1}, z_{j+1}, \ldots, z_n)$. The ring $\mathbb{F}(z^{(j)})[z_j]$ of polynomials in the single indeterminate z_j, with coefficients that are rational functions of the remaining variables, is again a ring of fractions of $\mathcal{P}$, with denominator set $\mathbb{F}[z^{(j)}] \setminus \{0\}$.

Proof. "only if": Let $s_i = \frac{p_i}{q_i}$ and $q := \gcd_{\mathcal{P}}(p_1, p_2) \in \mathcal{Q}$. As $\mathbb{F}(z^{(j)})[z_j]$ is a principal ideal domain, there exist $a_i^{(j)} \in \mathcal{P}$ and $b_i^{(j)} \in \mathbb{F}[z^{(j)}] \setminus \{0\}$ such that

$$\frac{a_1^{(j)}}{b_1^{(j)}} \cdot p_1 + \frac{a_2^{(j)}}{b_2^{(j)}} \cdot p_2 = q.$$

Hence

$$\frac{a_1^{(j)} b_2^{(j)} q_1}{q} \cdot \frac{p_1}{q_1} + \frac{a_2^{(j)} b_1^{(j)} q_2}{q} \cdot \frac{p_2}{q_2} = b_1^{(j)} b_2^{(j)}$$

which yields the desired representation.

"if": By (3.1), $\gcd(s_1, s_2)$ divides $p^{(j)}$ for all j. Then it also divides $\gcd(p^{(1)}, \ldots, p^{(n)})$. However, $\gcd_{\mathcal{P}}(p^{(1)}, \ldots, p^{(n)}) = 1$. $\quad\square$

The generalization of Lemma 3.1.2 to an arbitrary finite number of elements of S is obvious. Next, we consider co-primeness of matrices over S. For convenience, we sometimes suppress the matrix dimensions and simply write $M(S)$ for the set of matrices with entries in S (analogously for $\mathcal{R}$). By the rank of such a matrix, we mean its rank when considering it as an element of $M(\mathcal{R})$.

Definition 16 A full column rank matrix $M \in S^{l \times k}$ are said to be **minor right prime** if its minors of order k are co-prime. Two matrices $D, N \in M(S)$ with the same number of columns are called **minor right co-prime** if the block matrix $[N^T, D^T]^T$ is minor right prime. Left minor (co-)primeness is defined by transposition.

We derive a generalized matrix Bézout equation for co-prime matrices over S. The transition from the scalar to the matrix case is analogous to the case of co-primeness over $\mathcal{P}$ (see [88]), however, we give the proof for the sake of completeness.

Theorem 13 *A full column rank matrix $M \in S^{l \times k}$ is minor right prime iff there exists, for $1 \leq j \leq n$, a matrix $T^{(j)} \in S^{k \times l}$ such that*

$$T^{(j)} M = p^{(j)} I_k$$

where $0 \neq p^{(j)} \in P$ is independent of z_j.

Proof. "if": It follows from the Cauchy–Binet theorem [27] that

$$\det(T^{(j)} M) = \sum_i t_i^{(j)} m_i,$$

where m_i are the $k \times k$ minors of M, and $t_i^{(j)}$ are those of $T^{(j)}$. On the other hand, $\det(T^{(j)} M) = p^{(j)k}$. It follows from Lemma 3.1.2 that the minors of M must be co-prime.

"only if": For each minor m_i of M, there exists a permutation matrix Π_i such that $\Pi_i M = [M_{1i}^T, M_{2i}^T]^T$ with $\det(M_{1i}) = m_i$. Then $[\text{adj}\,(M_{1i}), 0]\Pi_i M = m_i I$, where $\text{adj}\,(\cdot)$ denotes the adjoint of a matrix. We conclude that, for each minor m_i, there exists $N_i \in S^{k \times l}$ such that $N_i M = m_i I$. Hence, if there exist $t_i^{(j)} \in S$ such that $\sum_i t_i^{(j)} m_i = p^{(j)}$, then $(\sum_i t_i^{(j)} N_i) M = p^{(j)} I$. □

Similarly, we have the following stronger co-primeness concept.

Definition 17 Two elements $s_1, s_2 \in S$ are called **zero co-prime** if there exist $t_1, t_2 \in S$ such that

$$t_1 s_1 + t_2 s_2 = 1,$$

that is, if the ideal generated by s_1 and s_2 is all of S. A matrix $M \in S^{l \times k}$ is said to be **zero right prime** if its minors of order k are zero co-prime, that is, if

$$\mathcal{I}_k(M) = S,$$

where $\mathcal{I}_k(M)$ denotes the ideal generated by the minors of M of order k. Two matrices $D, N \in M(S)$ with the same number of columns are called **zero right co-prime** if the block matrix $[N^T, D^T]^T$ is zero right prime. Zero left (co-) primeness is defined by transposition.

We give the following result without proof, as it is completely analogous to Theorem 13 above.

Corollary 12 *A matrix $M \in S^{l \times k}$ is zero right prime iff it possesses a left inverse over S, that is, if there exists a matrix $T \in S^{k \times l}$ such that*

$$TM = I_k.$$

Definition 18 We say that a rational matrix $H \in \mathcal{R}^{p \times m}$ has a minor (zero) right co-prime factorization if there exists $N \in \mathcal{S}^{p \times m}$, and a non-singular $D \in \mathcal{S}^{m \times m}$ such that

$$H = ND^{-1}$$

and D, N are minor (zero) right co-prime.

It turns out that these notions are restrictive in the sense that there are rational matrices that do not even permit a minor co-prime factorization over $\mathcal{S}$.

Example 3.1.2. Let $\mathcal{C}$ be the ring of causal rational functions, and let

$$H = \left[\begin{array}{cc} \frac{z_1}{z_3} & \frac{z_2}{z_3} \end{array} \right] \in \mathbb{R}(z_1, z_2, z_3)^{1 \times 2}.$$

Suppose that $H = ND^{-1}$ with $N \in \mathcal{C}^{1 \times 2}$ and $D \in \mathcal{C}^{2 \times 2}$. Then we must have

$$\left[\begin{array}{cc} z_1 & z_2 \end{array} \right] D = z_3 N.$$

This implies that all the entries of D must be multiples of z_3. But then the minors of $[N^T, D^T]^T$ of order 2 have $z_3 \notin U(\mathcal{C})$ as a common factor.

Our next goal is to get rid of the full rank requirement in Definition 16. We use the generalized notion of co-primeness from the previous chapter, which is such that any rational matrix possesses right and left co-prime factorizations in this sense.

Definition 19 A matrix $M \in M(\mathcal{S})$ is said to be **right prime** if, for any factorization $M = M_1 X$, where $M_1, X \in M(\mathcal{S})$, and rank$(M)$ = rank(M_1), we must have $M_1 = MY$ for some $Y \in M(\mathcal{S})$. Two matrices $N, D \in M(\mathcal{S})$ with the same number of columns are called **right co-prime** if the block matrix $[N^T, D^T]^T$ is right prime. Left (co-)primeness is defined by transposition.

Definition 20 A **right factorization** of $H \in M(\mathcal{R})$ is a pair of matrices $D, N \in M(\mathcal{S})$ such that $HD = N$ and D has full row rank. A right **co-prime** factorization is one in which D and N are right co-prime. Left (co-prime) factorizations are defined by transposition.

The following facts are known to hold for factorizations over $\mathcal{P}$ (see Theorem 8). Again, they translate straightforwardly to factorizations over $\mathcal{S}$, and we give proofs only for the sake of completeness.

Lemma 3.1.3. *A matrix $M \in M(\mathcal{S})$ is right prime iff it is a minimal right annihilator, that is, if there exists an $R \in M(\mathcal{S})$ whose kernel is generated by the columns of M.*

Proof. "only if": Let $M \in \mathcal{S}^{l \times k}$ be right prime, and let $R \in M(\mathcal{S})$ be a matrix such that $RM = 0$ and rank (R) + rank $(M) = l$. Such a matrix results, for instance, from computing a basis of the kernel of M^T over $\mathcal{R}$ (without loss of generality, such a basis can be chosen with entries in $\mathcal{S}$ only). Now let M_1 be minimal right annihilator of R, then rank (M) = rank (M_1) and

$$\ker(R) = \operatorname{im}(M_1) \supseteq \operatorname{im}(M). \tag{3.2}$$

But then there exists $X \in M(\mathcal{S})$ with $M = M_1 X$ and, due to the right primeness of M, $M_1 = MY$ for some $Y \in M(\mathcal{S})$. But then the three $\mathcal{S}$-modules in (3.2) are all equal, implying that also M is a minimal right annihilator (of R).

"if": Let M be a minimal right annihilator, say of R, and suppose that $M = M_1 X$ for some $X \in M(\mathcal{S})$ with rank (M) = rank (M_1). Then

$$\ker(R) = \operatorname{im}(M) \subseteq \operatorname{im}(M_1).$$

Certainly, $M_1 = M\tilde{Y}$ for some $\tilde{Y} \in M(\mathcal{R})$. Thus $RM_1 = 0$, showing that the three modules are again equal. Hence we must have $M_1 = MY$ for some $Y \in M(\mathcal{S})$. $\qquad \square$

Theorem 14 *Any rational matrix possesses right and left co-prime factorizations.*

Proof. Let $H \in \mathcal{R}^{p \times m}$ be given. There exists a natural representation $H = \bar{N}/d$ with $\bar{N} \in \mathcal{S}^{p \times m}$ and $0 \neq d \in \mathcal{S}$. Thus, with $\bar{D} := dI_p$ we have $H = \bar{D}^{-1}\bar{N}$. Now define the block matrix $R := [-\bar{D}, \bar{N}]$ and compute a minimal right annihilator of R, say $M = [N^T, D^T]^T$ with the suitable partition. Then D, N provides a right co-prime factorization of H, because $N = \bar{D}^{-1}\bar{N}D = HD$ and

$$\operatorname{rank}(D) = \operatorname{rank} \begin{bmatrix} HD \\ D \end{bmatrix} = \operatorname{rank} \begin{bmatrix} N \\ D \end{bmatrix} = p + m - \operatorname{rank} \begin{bmatrix} -\bar{D} & \bar{N} \end{bmatrix} = m.$$

$\qquad \square$

Note that if D, N and D_1, N_1 are two right co-prime factorizations of H, then

$$\mathcal{M} := \operatorname{im} \begin{bmatrix} N \\ D \end{bmatrix} = \operatorname{im} \begin{bmatrix} N_1 \\ D_1 \end{bmatrix}.$$

Thus, the matrix $H \in \mathcal{R}^{p \times m}$ determines a finitely generated module $\mathcal{M} \subset \mathcal{S}^{p+m}$. Conversely, any generating system M of $\mathcal{M} = \operatorname{im}(M)$ gives rise to right co-prime factorizations (by selecting m rows that are linearly independent over $\mathcal{R}$ and collecting them, possibly after a permutation of rows, in the "denominator" matrix D). Consider the module $\mathcal{W} := \mathcal{S}^{p+m}/\mathcal{M}$ and its dual counterpart $\mathcal{T}$ (constructed from left rather than right co-prime factorizations).

Lemma 3.1.4. *[73] A rational matrix possesses a zero right and a zero left co-prime factorization iff the associated modules $\mathcal{W}$ and $\mathcal{T}$ as constructed above are free.*

Proof. Let D, N be a right co-prime factorization of $H \in \mathcal{R}^{p \times m}$ and let

$$M = \begin{bmatrix} N \\ D \end{bmatrix} \in \mathcal{S}^{(p+m) \times k}.$$

Suppose that $\mathcal{W} = \text{cok}(M)$ is free. Noting that $\text{rank}\,(M) = m$, this implies that

$$\text{cok}(M) = \mathcal{S}^{p+m}/\text{im}(M) \cong \mathcal{S}^p.$$

Then there exists $K \in \mathcal{S}^{p \times (p+m)}$ such that

$$\mathcal{S}^k \xrightarrow{M} \mathcal{S}^{p+m} \xrightarrow{K} \mathcal{S}^p \to 0$$

is an exact sequence. This means that M is a minimal right annihilator of K, which is surjective, that is, zero left prime. With the partition $K = [-\bar{D}, \bar{N}]$, where $\bar{D} \in \mathcal{S}^{p \times p}$, $\bar{N} \in \mathcal{S}^{p \times m}$, we have $(\bar{D}P - \bar{N})D_P = 0$, and since D_P has full row rank, $\bar{D}P = \bar{N}$. Finally, as $\text{rank}\,(K) = p = \text{rank}\,(\bar{D})$, we have $P = \bar{D}^{-1}\bar{N}$, which constitutes a zero left co-prime factorization of P. Similarly, one shows that H has a zero right co-prime factorization if $\mathcal{T}$ is free.

For the converse, let $H = \bar{N}/d$ be any factorization of H over $\mathcal{S}$, with a scalar d. Let D, N be a right co-prime factorization of H, then

$$M = \begin{bmatrix} N \\ D \end{bmatrix}$$

is a minimal right annihilator of

$$W = \begin{bmatrix} dI_p & -\bar{N} \end{bmatrix},$$

that is, $\ker(W) = \text{im}(M)$. Consider the mapping

$$W: \quad \mathcal{S}^{p+m} \to \mathcal{S}^p, \quad x \mapsto Wx.$$

Then

$$\mathcal{W} = \mathcal{S}^{p+m}/\text{im}(M) = \mathcal{S}^{p+m}/\ker(W) \cong \text{im}(W)$$

by the homomorphism theorem. Now if H has zero left and right factorizations, say $H = AB^{-1} = \bar{B}^{-1}\bar{A}$ with Bézout equations $XA + YB = I$ and $\bar{A}\bar{X} + \bar{B}\bar{Y} = I$, we have

$$\begin{bmatrix} X & Y \\ -\bar{B} & \bar{A} \end{bmatrix} \begin{bmatrix} A & -\bar{Y} \\ B & \bar{X} \end{bmatrix} = \begin{bmatrix} I & * \\ 0 & I \end{bmatrix}.$$

The matrix on the right hand side is unimodular (*i.e.*, it possesses an inverse in $M(\mathcal{S})$), and thus the matrices on the left hand side must also be unimodular. Thus

$$\left[\begin{array}{cc} dI_p & -\bar{N} \end{array}\right] \left[\begin{array}{cc} A & -\bar{Y} \\ B & \bar{X} \end{array}\right] = \left[\begin{array}{cc} 0 & -Z \end{array}\right],$$

where $Z = d\bar{Y} + \bar{N}\bar{X}$ is a non-singular $p \times p$ matrix with $\mathrm{im}(W) = \mathrm{im}(Z)$. This shows that W is free. A similar argument applies to $\mathcal{T}$.   □

The subsequent section gives sufficient and necessary conditions for stabilizability of a rational matrix in terms of the associated modules W and $\mathcal{T}$.

3.2 Feedback Stabilization

As signals, we consider n-fold indexed sequences with values in $\mathbb{F}$, that is, our signal space is $\mathcal{A} = \mathbb{F}^{N^n}$. A causal plant $H \in \mathcal{C}^{p \times m}$ has a formal power series representation

$$H = \sum_{l_1=0}^{\infty} \cdots \sum_{l_n=0}^{\infty} h(l_1, \ldots, l_n) z_1^{l_1} \cdots z_n^{l_n}$$

where $h(l_1, \ldots, l_n) \in \mathbb{F}^{p \times m}$. Hence, given an input signal $u \in \mathcal{A}^m$, we define the corresponding output $y = Hu$ by

$$y(k_1, \ldots, k_n) = \sum_{l_1=0}^{k_1} \cdots \sum_{l_n=0}^{k_n} h(l_1, \ldots, l_n) u(k_1 - l_1, \ldots, k_n - l_n)$$

where $k_j \in \mathbb{N}$. Consider the following standard feedback configuration:

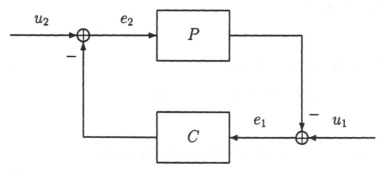

Fig. 3.1. Feedback Loop.

The system equations are

$$\left[\begin{array}{cc} I & P \\ C & I \end{array}\right] \left[\begin{array}{c} e_1 \\ e_2 \end{array}\right] = \left[\begin{array}{c} u_1 \\ u_2 \end{array}\right]. \tag{3.3}$$

Definition 21 We call the feedback system **well-posed** if the rational matrix appearing in (3.3) is non-singular, which is true iff $P \in \mathcal{R}^{p \times m}$ and $C \in \mathcal{R}^{m \times p}$ are such that $\det(I - PC) \neq 0$. We say that C is a **stabilizing controller** for P (or, for short, C **stabilizes** P) if

$$H(P,C) := \begin{bmatrix} I & P \\ C & I \end{bmatrix}^{-1} \in M(\mathcal{S}). \tag{3.4}$$

A plant P is called **stabilizable** if it possesses a stabilizing controller C.

Although an action of non-causal rational functions on signals has not been defined, we consider the stabilization problem for arbitrary rational matrices P and C. Of course, we might just as well restrict to the ring $\mathcal{C}$ of causal rational functions, requiring that $\mathcal{P} \subseteq \mathcal{S} \subseteq \mathcal{C} \subseteq \mathcal{R}$.

Our next goal is to characterize the stabilization property (3.4) in terms of right co-prime factorizations of plant and controller, in a well-posed loop. For this, let $PD_P = N_P$ and $CD_C = N_C$ be right co-prime factorizations of P and C, respectively. Then we have

$$\begin{bmatrix} I & P \\ C & I \end{bmatrix} \begin{bmatrix} D_C & 0 \\ 0 & D_P \end{bmatrix} = \begin{bmatrix} D_C & N_P \\ N_C & D_P \end{bmatrix}.$$

Since D_C and D_P have full row rank, and since the leftmost matrix is non-singular, the matrix on the right hand side has full row rank. Hence

$$\begin{bmatrix} D_C & 0 \\ 0 & D_P \end{bmatrix} = H(P,C) \begin{bmatrix} D_C & N_P \\ N_C & D_P \end{bmatrix} \tag{3.5}$$

is a right factorization of $H(P,C)$.

Lemma 3.2.1. *The factorization (3.5) is right co-prime.*

Proof. We need to show that

$$\begin{bmatrix} D_C & 0 \\ 0 & D_P \\ D_C & N_P \\ N_C & D_P \end{bmatrix} = \begin{bmatrix} I & 0 & 0 & 0 \\ 0 & I & 0 & 0 \\ I & 0 & I & 0 \\ 0 & I & 0 & I \end{bmatrix} \begin{bmatrix} D_C & 0 \\ 0 & D_P \\ 0 & N_P \\ N_C & 0 \end{bmatrix}$$

is right prime. However, this follows directly from the right co-primeness of D_C, N_C and D_P, N_P, respectively. $\square$

Theorem 15 Let $D, N \in M(\mathcal{S})$ be a right co-prime factorization of $H \in M(\mathcal{R})$. We have $H \in M(\mathcal{S})$ iff there exists a right inverse $D^b \in M(\mathcal{S})$ of D.

Proof. "if": We have $HD = N$. If $DD^b = I$ for some $D^b \in M(\mathcal{S})$, then $H = ND^b \in M(\mathcal{S})$.

"only if": Suppose that $H \in \mathcal{S}^{p \times m}$, then $\text{rank}(D) = m$ and

$$\begin{bmatrix} N \\ D \end{bmatrix} = \begin{bmatrix} H \\ I_m \end{bmatrix} D$$

is a factorization of $[N^T, D^T]^T$ over S. We have

$$\operatorname{rank} \begin{bmatrix} N \\ D \end{bmatrix} = \operatorname{rank} \begin{bmatrix} HD \\ D \end{bmatrix} = \operatorname{rank}(D) = m = \operatorname{rank} \begin{bmatrix} H \\ I_m \end{bmatrix}.$$

The right co-primeness of D and N implies that there exists $D^b \in M(S)$ such that

$$\begin{bmatrix} H \\ I_m \end{bmatrix} = \begin{bmatrix} N \\ D \end{bmatrix} D^b.$$

$\square$

Corollary 13 *Let P and C be such that $I - PC$ is non-singular. Let $PD_P = N_P$ and $CD_C = N_C$ be right co-prime factorizations of P and C, respectively. Then C stabilizes P iff*

$$R := \begin{bmatrix} D_C & N_P \\ N_C & D_P \end{bmatrix} \qquad (3.6)$$

possesses a right inverse over S, that is, iff R is zero left prime.

A necessary condition for the existence of a stabilizing controller is now easily obtained: Suppose that R from (3.6) is zero left prime, that is,

$$\mathcal{I}_{p+m}(R) = S. \qquad (3.7)$$

Consider the minors of R of order $p + m$, each corresponding to a selection of $p + m$ columns of $R = [M_C, M_P]$. As $\operatorname{rank}(M_C) = p$ and $\operatorname{rank}(M_P) = m$, we must take precisely p columns of M_C and m columns of M_P, in order to obtain a non-zero element of $\mathcal{I}_{p+m}(R)$. Using determinant expansion by minors, this implies

$$\mathcal{I}_{p+m}(R) \subseteq \mathcal{I}_p(M_C) \cap \mathcal{I}_m(M_P).$$

Combining this with (3.7) we get $\mathcal{I}_p(M_C) = S = \mathcal{I}_m(M_P)$ as a necessary condition for $R = [M_C, M_P]$ to be zero left prime.

Theorem 16

1. *If P admits zero right and left co-prime factorizations, that is, if the modules W and T are free, then P is stabilizable.*
2. *If P is stabilizable, then W and T are projective.*

Proof. 1. If P admits zero right and left co-prime factorizations, the set of stabilizing controllers for P can be parameterized like in the 1D case [87]; see also [76], and for the multidimensional case [82]. To see that the set is indeed non-empty, one may employ an argument given by Vidyasagar in [76, p. 96].

2. Let D_P, N_P be a right co-prime factorization of P, that is,

$$M = \begin{bmatrix} N_P \\ D_P \end{bmatrix} \in \mathcal{S}^{(p+m) \times k}$$

is a right prime matrix with $\text{rank}(M) = m$ and $PD_P = N_P$. If there exists a stabilizing controller for P, we must have $\mathcal{I}_m(M) = \mathcal{S}$ according to the remark following Corollary 13. Then $\mathcal{W} = \text{cok}(M)$ is projective [13] and a similar argument applies to $\mathcal{T}$.

□

The preceding theorem characterizes stabilizability as a property that is weaker than freeness, and stronger than projectivity, of certain $\mathcal{S}$-modules associated to a given plant, where $\mathcal{S}$ denotes a ring of stable transfer functions. The question over which rings projective modules are already free, is one of the fundamental problems of algebra, see for instance Lam's monograph [40] on Serre's conjecture.

4. Strict System Equivalence

The notion of strict system equivalence was introduced by Rosenbrock [63] and was subsequently refined by Fuhrmann [25]. It describes the connection between all least order realizations of a transfer matrix in terms of a polynomial system

$$Tx = Uu \tag{4.1}$$

$$y = Vx + Wu. \tag{4.2}$$

Pugh *et al.* [57] generalized the notion to two-dimensional systems. The aim of this chapter is to give an account of the general r-dimensional situation. As pointed out in the previous chapter, the main difficulty is that the usual concept of a right co-prime factorization of a rational matrix, which is of the form $G = ND^{-1}$ with a polynomial matrix N, and a square non-singular matrix D, is restrictive for systems of dimension $r > 2$. Non-square "denominator" matrices D have to be admitted, and consequently, rectangular T-matrices have to be considered in (4.1).

Let $\mathbb{F}$ be an arbitrary field, and let G be a rational $p \times m$ matrix in r indeterminates, with coefficients in $\mathbb{F}$, that is,

$$G \in \mathbb{F}(s_1, \dots, s_r)^{p \times m}.$$

As usual, the term "polynomial matrix" will refer to matrices with entries in the polynomial ring $\mathbb{F}[s_1, \dots, s_r]$, and $\mathcal{A}$ denotes one of the usual continuous or discrete signal spaces.

Let D, N be a right factorization of G, that is, D and N are polynomial matrices with $GD = N$, and D has full row rank. Then the system of linear constant-coefficient partial differential or difference equations, for $u \in \mathcal{A}^m$, $x \in \mathcal{A}^n$, $y \in \mathcal{A}^p$,

$$Dx = u \tag{4.3}$$

$$y = Nx \tag{4.4}$$

is called a **realization** of G. These are precisely the driving-variable realizations discussed in Chapter 2. A more general version of (4.3), (4.4) is given by

$$Tx = Uu$$
$$y = Vx + Wu$$

where T, U, V, W are polynomial matrices of dimensions $\rho \times n$, $\rho \times m$, $p \times n$, $p \times m$, respectively, T has full row rank, and $V = HT$ for some rational matrix H. The transfer function is $G = HU + W$ as

$$Gu = (HU + W)u = HTx + Wu = Vx + Wu = y.$$

This constitutes the class of systems we are going to consider.

Definition 22 *An* **admissible** *system*

$$\Sigma = (T, U, V, W)$$

is given by polynomial matrices T, U, V, W of size $\rho \times n$, $\rho \times m$, $p \times n$, $p \times m$, respectively, with the following properties:

1. *T has full row rank;*
2. *there exists a rational matrix H such that $V = HT$.*

The rational matrix $G = HU + W$ is called the **transfer function** *of Σ, and Σ is said to be a* **realization** *of G.*

We will also use the associated system matrix

$$P = \begin{bmatrix} T & U \\ -V & W \end{bmatrix} = \begin{bmatrix} T & U \\ -HT & W \end{bmatrix}.$$

As T has full row rank, there exists a *rational* matrix X such that $TX = I_\rho$. Hence $VX = HTX = H$ and $G = HU + W = VXU + W$. Note that the existence of H with $V = HT$ has to be postulated in our definition of an admissible system ($H = VT^{-1}$ in the familiar case when T is square and non-singular), whereas the dual property, $U = TK$ for some rational matrix K, is already implied by the rank requirement on T (take $K = XU$).

Example 4.0.1. Consider once more the second set of Maxwell equations (1.33), (1.34)

$$\frac{\partial \boldsymbol{D}}{\partial t} = \nabla \times \boldsymbol{H} - \boldsymbol{J}$$
$$\rho = \nabla \cdot \boldsymbol{D}.$$

Recall that this is a full row rank system of four equations for ten unknown functions. Thus there are six free variables, e.g., the pairs $(\boldsymbol{D}, \boldsymbol{H})$ or $(\boldsymbol{H}, \boldsymbol{J})$ can be considered as free variables of the system (but not $(\boldsymbol{D}, \boldsymbol{J})$). Although this cause—effect interpretation may be counter-intuitive from a physicist's point of view, it is mathematically justified, by the nature of this set of equations. We take $u = (\boldsymbol{H}, \boldsymbol{J})$, $x = \boldsymbol{D}$, and $y = \rho$, that is, $T = s_4 I_3$, $U = [\text{ curl}, -I_3]$, $V = \text{div}$, $W = 0$, where

$$\text{curl} = \begin{bmatrix} 0 & -s_3 & s_2 \\ s_3 & 0 & -s_1 \\ -s_2 & s_1 & 0 \end{bmatrix} \quad \text{and} \quad \text{div} = [\ s_1 \ \ s_2 \ \ s_3\ ].$$

It is easy to see that this system is admissible with $H = \frac{1}{s_4}\text{div}$ and transfer function

$$G = HU = -\frac{1}{s_4}[\ 0, \ \ \text{div}\ ] \in \mathbb{R}(s_1, s_2, s_3, s_4)^{1 \times 6}.$$

Another interpretation, which is even of type (4.3), (4.4), is indicated by

$$\nabla \times \boldsymbol{H} - \frac{\partial \boldsymbol{D}}{\partial t} = \boldsymbol{J}$$
$$\rho = \nabla \cdot \boldsymbol{D},$$

that is, $T = [\text{ curl}, -s_4 I_3]$, $U = I$, $V = [\ 0, \ \ \text{div}\ ]$, $W = 0$. Its transfer matrix is

$$G = -\frac{1}{s_4}\text{div}$$

and T, V are even a right co-prime factorization of G, that is, the system is canonical in a sense to be defined below.

4.1 System Equivalence

Definition 23 *Two admissible systems* (T_i, U_i, V_i, W_i) *are said to be* **system equivalent** *if there exist rational matrices* Q_l, R_l, Q_r, R_r *of dimensions* $p_2 \times p_1$, $p \times p_1$, $n_2 \times n_1$, $n_2 \times m$, *respectively, such that*

$$\begin{bmatrix} Q_l & 0 \\ R_l & I_p \end{bmatrix}\begin{bmatrix} T_1 & U_1 \\ -V_1 & W_1 \end{bmatrix} = \begin{bmatrix} T_2 & U_2 \\ -V_2 & W_2 \end{bmatrix}\begin{bmatrix} Q_r & R_r \\ 0 & I_m \end{bmatrix}. \tag{4.5}$$

Theorem 17 *Two admissible systems* $(T_i, U_i, V_i = H_i T_i, W_i)$ *are realizations of the same transfer function iff they are system equivalent.*

Proof. "if": The constituent equations of (4.5) are

$$Q_l T_1 = T_2 Q_r \tag{4.6}$$
$$Q_l U_1 = T_2 R_r + U_2 \tag{4.7}$$
$$R_l T_1 - V_1 = -V_2 Q_r \tag{4.8}$$
$$R_l U_1 + W_1 = -V_2 R_r + W_2 \tag{4.9}$$

Using (4.7),

$$G_2 = H_2 U_2 + W_2 = H_2(Q_l U_1 - T_2 R_r) + W_2 = H_2 Q_l U_1 - V_2 R_r + W_2.$$

With (4.9), it follows that

$$G_2 = H_2 Q_l U_1 + R_l U_1 + W_1 = (H_2 Q_l + R_l)U_1 + W_1.$$

It remains to be shown that $H_2 Q_l + R_l = H_1$. To see this, consider

$$V_2 Q_r = H_2 T_2 Q_r = H_2 Q_l T_1, \tag{4.10}$$

where the second equality holds due to (4.6). Finally, we use (4.8) to obtain

$$V_2 Q_r = -R_l T_1 + V_1 = -R_l T_1 + H_1 T_1. \tag{4.11}$$

Comparing (4.10) and (4.11) yields $(H_2 Q_l + R_l)T_1 = H_1 T_1$. Using that T_1 has full row rank, we conclude that $H_2 Q_l + R_l = H_1$ as desired.

"only if": Suppose that $G_1 = H_1 U_1 + W_1 = H_2 U_2 + W_2 = G_2$. Let X_i be such that $T_i X_i = I_{\rho_i}$. Then $V_i X_i = H_i$ and

$$\begin{bmatrix} FH_1 & 0 \\ H_1 - H_2 FH_1 & I_p \end{bmatrix} \begin{bmatrix} T_1 & U_1 \\ -V_1 & W_1 \end{bmatrix} =$$

$$= \begin{bmatrix} T_2 & U_2 \\ -V_2 & W_2 \end{bmatrix} \begin{bmatrix} X_2 FV_1 & X_2(FH_1 U_1 - U_2) \\ 0 & I_m \end{bmatrix}$$

holds for any rational $\rho_2 \times p$ matrix F. We prove the claimed equality componentwise. For the first row, we use $T_2 X_2 = I$, hence

$$T_2 X_2 FV_1 = FV_1 = FH_1 T_1$$

and

$$T_2 X_2(FH_1 U_1 - U_2) + U_2 = FH_1 U_1 - U_2 + U_2 = FH_1 U_1.$$

For the (2,1)-entry, we have

$$(H_1 - H_2 FH_1)T_1 - V_1 = -H_2 FH_1 T_1 = -H_2 FV_1 = -V_2 X_2 FV_1.$$

Finally, we use the equality of the transfer functions in

$$\begin{aligned} (H_1 - H_2 FH_1)U_1 + W_1 &= H_1 U_1 - H_2 FH_1 U_1 + W_1 \\ &= H_2 U_2 - H_2 FH_1 U_1 + W_2 \\ &= -H_2(FH_1 U_1 - U_2) + W_2 \\ &= -V_2 X_2(FH_1 U_1 - U_2) + W_2. \end{aligned}$$

$\square$

In particular, Theorem 17 implies that system equivalence is indeed an equivalence relation on the set of admissible systems. Equation (4.5) with *polynomial* intertwining matrices defines a relation between the system matrices P_1 and P_2 which is reflexive and transitive. Additional constraints have to be imposed to make the relation symmetric. Rosenbrock [63] proposed unimodularity of Q_l and Q_r, but this requires the system matrices to be of the same size ($\rho_1 = \rho_2$ and $n_1 = n_2$). Fuhrmann [25] demanded that Q_l, T_2 be left co-prime and T_1, Q_r right co-prime. Even if we use the strongest multivariate co-primeness notion, zero co-primeness, when adapting Fuhrmann's notion to the multivariate case, we still need that $\rho_1 + n_2 = \rho_2 + n_1$ to have symmetry of the relation.

Example 4.1.1. Consider

$$\begin{bmatrix} 1 & 0 \\ 0 & 1 \end{bmatrix} \begin{bmatrix} t_1+t_2 & u \\ -v_1-v_2 & w \end{bmatrix} = \begin{bmatrix} t_1 & t_2 & u \\ -v_1 & -v_2 & w \end{bmatrix} \begin{bmatrix} 1 & 0 \\ 1 & 0 \\ 0 & 1 \end{bmatrix}$$

where t_1, t_2, u, v_1, v_2, w are scalar polynomials. The converse relation takes the form

$$\begin{bmatrix} q_l & 0 \\ r_l & 1 \end{bmatrix} \begin{bmatrix} t_1 & t_2 & u \\ -v_1 & -v_2 & w \end{bmatrix} = \begin{bmatrix} t_1+t_2 & u \\ -v_1-v_2 & w \end{bmatrix} \begin{bmatrix} q_{r1} & q_{r2} & r_r \\ 0 & 0 & 1 \end{bmatrix}.$$

As $q_l \begin{bmatrix} t_1, & t_2 \end{bmatrix} = (t_1 + t_2) \begin{bmatrix} q_{r1}, & q_{r2} \end{bmatrix}$, it follows that the zero right co-primeness requirement on $T_1 = \begin{bmatrix} t_1, & t_2 \end{bmatrix}$ and $Q_r = \begin{bmatrix} q_{r1}, & q_{r2} \end{bmatrix}$ cannot be satisfied, as the two row vectors are rationally linearly dependent.

Section 4.3 will yield an interpretation of the requirement $\rho_1 + n_2 = \rho_2 + n_1$. The trivial way to obtain an equivalence relation $\approx$ from a reflexive and transitive relation $\sim$ is to define

$$P_1 \approx P_2 \quad :\Leftrightarrow \quad P_1 \sim P_2 \text{ and } P_2 \sim P_1$$

and we will use this in our definition of strict system equivalence below.

4.2 Strict System Equivalence

Definition 24 *We say that two admissible systems $\Sigma_i = (T_i, U_i, V_i = H_i T_i, W_i)$ of the same size ($\rho_1 = \rho_2$, $n_1 = n_2$) are **strictly system equivalent in the Rosenbrock sense** (RSSE) if (4.5) is true with polynomial intertwining matrices and Q_l, Q_r being unimodular.*

*For admissible systems with $\rho_1 + n_2 = \rho_2 + n_1$, we define **strict system equivalence in the Fuhrmann sense** (FSSE) by equation (4.5) with polynomial matrices such that Q_l, T_2 are zero left co-prime and T_1, Q_r are zero right co-prime.*

*Two admissible systems of arbitrary size are said to be **strictly system equivalent (SSE)** if there are polynomial matrices $Q_l, Q_r, \tilde{Q}_l, \tilde{Q}_r$ and $R_l, R_r, \tilde{R}_l, \tilde{R}_r$ such that*

$$\begin{bmatrix} Q_l & 0 \\ R_l & I_p \end{bmatrix} \begin{bmatrix} T_1 & U_1 \\ -V_1 & W_1 \end{bmatrix} = \begin{bmatrix} T_2 & U_2 \\ -V_2 & W_2 \end{bmatrix} \begin{bmatrix} Q_r & R_r \\ 0 & I_m \end{bmatrix}$$

and

$$\begin{bmatrix} \tilde{Q}_l & 0 \\ \tilde{R}_l & I_p \end{bmatrix} \begin{bmatrix} T_2 & U_2 \\ -V_2 & W_2 \end{bmatrix} = \begin{bmatrix} T_1 & U_1 \\ -V_1 & W_1 \end{bmatrix} \begin{bmatrix} \tilde{Q}_r & \tilde{R}_r \\ 0 & I_m \end{bmatrix}.$$

The following facts can be proven analogously to the one-dimensional case. The first one also follows from Theorem 20 below. A special case of the second can be found in [57], and the proof given there translates straightforwardly to the present situation. For the sake of completeness, we give it in an appendix at the end of this chapter.

1. FSSE is an equivalence relation. Thus clearly, RSSE $\Rightarrow$ FSSE $\Rightarrow$ SSE.
2. Two admissible systems with $\rho_1 + n_2 = \rho_2 + n_1$ are FSSE iff a trivial expansion of one system is RSSE to a trivial expansion of the other. By a trivial expansion of (T, U, V, W), we mean a system with system matrix

$$\left[\begin{array}{c|cc} I & 0 & 0 \\ \hline 0 & T & U \\ 0 & -V & W \end{array}\right].$$

   The size of the identity blocks is chosen such that the expanded systems have T-matrices of the same size (note that this is only possible if $\rho_1 + n_2 = \rho_2 + n_1$ holds).
3. As an immediate consequence, we have that FSSE preserves left (zero) co-primeness of the T-, U-matrices and right (zero) co-primeness of the T-, V-matrices.

Definition 25 *By the **determinantal ideal** of an admissible system*

$$\Sigma = (T, U, V, W),$$

we mean the determinantal ideal of T, that is, the ideal generated by the minors of T of maximal order (i.e., of order $\rho \times \rho$, recalling that T has full row rank).

The following fact on determinantal ideals is fundamental [47, p. 7]: If two $\rho \times n$ matrices T_1 and T_2 are equivalent, *i.e.*, if there exist unimodular matrices Q_l, Q_r such that $Q_l T_1 = T_2 Q_r$, then we have $\mathcal{I}(T_1) = \mathcal{I}(T_2)$. Moreover, a trivial expansion does not change the determinantal ideal. The following theorem is an immediate consequence.

Theorem 18 *Two systems that are FSSE give rise to the same transfer matrix and have the same determinantal ideal.*

This generalizes a theorem of Rosenbrock [63, p. 52]. For a square non-singular matrix T,

$$\mathcal{I}(T) = \langle \det(T) \rangle,$$

the principal ideal generated by the determinant of T. Then equality of the determinantal ideals means nothing but equality of determinants (up to a non-zero constant factor).

Definition 26 *An admissible system $\Sigma = (T, U, V, W)$ is said to be **strongly controllable** (SC) if T, U are zero left co-prime. An SC system is called **canonical** if T and V are generalized factor right co-prime.*

For $\mathbb{F} = \mathbb{R}$ or $\mathbb{F} = \mathbb{C}$, strong controllability is closely related to Rosenbrock's input-decoupling zeros [63, p. 64]. In fact, zero left co-primeness of T, U is equivalent to

$$\text{rank} \left[\begin{array}{cc} T(\beta) & U(\beta) \end{array} \right] = \text{rank} \left[\begin{array}{cc} T & U \end{array} \right] = \rho \quad \text{for all } \beta \in \mathbb{C}^r.$$

At first sight, the notion of a canonical system may seem restrictive, but recall from (4.3), (4.4) that any transfer function possesses canonical realizations, since any rational matrix admits right co-prime factorizations. Canonical systems are "minimal" in the following sense (see Corollary 14 below): If an SC system (T_1, U_1, V_1, W_1) and a canonical system (T, U, V, W) give rise to the same transfer function, then

$$\mathcal{I}(T_1) \subseteq \mathcal{I}(T).$$

The corresponding relation for square non-singular T-matrices is that $\det(T)$ divides $\det(T_1)$. Hence a canonical system is a system of least order [63] in that sense (among the set of all SC realizations of a transfer function).

Theorem 19 *An SC system $\Sigma = (T, U, V = HT, W)$ is FSSE to an admissible system of the form $\widetilde{\Sigma} = (\widetilde{T}, I, \widetilde{V} = G\widetilde{T}, 0)$, where $G = HU + W$.*

Proof. Let the size of T and U be $\rho \times n$ and $\rho \times m$, respectively. As T, U are zero left co-prime, they can be incorporated into the first ρ rows of a unimodular matrix [89], that is, there exist polynomial matrices Q_r, R_r such that

$$\left[\begin{array}{cc} T & U \\ Q_r & R_r \end{array} \right]$$

is unimodular. Thus there exist polynomial matrices $Y, Z, Q_l, \widetilde{T}$ of dimensions $n \times \rho$, $n \times \tilde{n}$, $m \times \rho$, $m \times \tilde{n}$, respectively, where $\tilde{n} := n + m - \rho$, such that

$$\begin{bmatrix} Y & Z \\ Q_l & -\tilde{T} \end{bmatrix} = \begin{bmatrix} T & U \\ Q_r & R_r \end{bmatrix}^{-1}. \tag{4.12}$$

Hence

$$\begin{bmatrix} 0 & I \\ V & -W \end{bmatrix} \begin{bmatrix} Y & Z \\ Q_l & -\tilde{T} \end{bmatrix} \begin{bmatrix} T & U \\ Q_r & R_r \end{bmatrix} = \begin{bmatrix} 0 & I \\ V & -W \end{bmatrix}$$

which yields, with $R_l := VY - WQ_l$ and $\tilde{V} := VZ + W\tilde{T}$,

$$\begin{bmatrix} Q_l & -\tilde{T} \\ R_l & \tilde{V} \end{bmatrix} \begin{bmatrix} T & U \\ Q_r & R_r \end{bmatrix} = \begin{bmatrix} 0 & I \\ V & -W \end{bmatrix}.$$

This can be rewritten as

$$\begin{bmatrix} Q_l & 0 \\ R_l & I \end{bmatrix} \begin{bmatrix} T & U \\ -V & W \end{bmatrix} = \begin{bmatrix} \tilde{T} & I \\ -\tilde{V} & 0 \end{bmatrix} \begin{bmatrix} Q_r & R_r \\ 0 & I \end{bmatrix}.$$

Noting that the sizes of T and $\tilde{T}$ are compatible for FSSE, this establishes the desired relation, since the left zero co-primeness of $Q_l, \tilde{T}$ and the right zero co-primeness of T, Q_r are direct consequences of (4.12). It remains to be shown that $\tilde{\Sigma} = (\tilde{T}, I, \tilde{V}, 0)$ is an admissible system. From (4.12), $TZ = U\tilde{T}$, hence

$$\tilde{V} = VZ + W\tilde{T} = HTZ + W\tilde{T} = (HU + W)\tilde{T} = G\tilde{T}$$

with $G = HU + W$ as desired. Finally, assume that the first ρ columns of T are linearly independent (otherwise, perform a suitable permutation). In view of (4.12), a Schur complement argument (see the appendix at the end of this chapter) shows that

$$\begin{bmatrix} Z_2 \\ -\tilde{T} \end{bmatrix}$$

must be invertible, where Z_2 denotes the last $n - \rho$ rows of Z. In particular, $\tilde{T}$ has full row rank.   □

Corollary 14 *If an SC system $\Sigma_1 = (T_1, U_1, V_1, W_1)$ and a canonical system $\Sigma = (T, U, V, W)$ give rise to the same transfer function, then*

$$\mathcal{I}(T_1) \subseteq \mathcal{I}(T).$$

In particular, two canonical realizations of a transfer matrix have the same determinantal ideal.

Proof. By Theorem 19, the systems can be reduced by FSSE to the respective forms $\tilde{\Sigma}_1 = (\tilde{T}_1, I, \tilde{V}_1 = G\tilde{T}_1, 0)$ and $\tilde{\Sigma} = (\tilde{T}, I, \tilde{V} = G\tilde{T}, 0)$. As FSSE preserves the co-primeness of T, V, it follows that $\tilde{T}_1, \tilde{V}_1$ and $\tilde{T}, \tilde{V}$ are two right factorizations of G with $\tilde{T}, \tilde{V}$ being even right co-prime. But this implies that there exists a polynomial matrix Y such that

$$\left[\begin{array}{c} \tilde{T}_1 \\ \tilde{V}_1 \end{array}\right] = \left[\begin{array}{c} \tilde{T} \\ \tilde{V} \end{array}\right] Y.$$

As rank $(\tilde{T}) = $ rank $(\tilde{T}_1) = m$, we conclude [47, p. 7] that $\mathcal{I}(\tilde{T}_1) = \mathcal{I}(\tilde{T}Y) \subseteq \mathcal{I}(\tilde{T})$. According to Theorem 18, we have $\mathcal{I}(T) = \mathcal{I}(\tilde{T})$ and $\mathcal{I}(T_1) = \mathcal{I}(\tilde{T}_1)$. This yields the desired result. $\square$

The following corollary should be compared to the corresponding result of Rosenbrock [63, p. 106].

Corollary 15 *Two canonical systems are SSE iff they are realizations of the same transfer function.*

Proof. Necessity follows directly from Theorem 17. For sufficiency, let

$$\Sigma_i = (T_i, U_i, V_i = H_i T_i, W_i)$$

be canonical systems with $G := H_1 U_1 + W_1 = H_2 U_2 + W_2$. By Theorem 19, they can be reduced by FSSE to the respective forms $\tilde{\Sigma}_i = (\tilde{T}_i, I, \tilde{V}_i = G\tilde{T}_i, 0)$. As FSSE preserves the co-primeness of T_i, V_i, it follows that $\tilde{T}_1$, $\tilde{V}_1$ and $\tilde{T}_2, \tilde{V}_2$ are two right co-prime factorizations of G. Hence there exist polynomial matrices Y and Z such that

$$\left[\begin{array}{cc} \tilde{T}_1 & I \\ -\tilde{V}_1 & 0 \end{array}\right] = \left[\begin{array}{cc} \tilde{T}_2 & I \\ -\tilde{V}_2 & 0 \end{array}\right] \left[\begin{array}{cc} Y & 0 \\ 0 & I \end{array}\right]$$

and

$$\left[\begin{array}{cc} \tilde{T}_2 & I \\ -\tilde{V}_2 & 0 \end{array}\right] = \left[\begin{array}{cc} \tilde{T}_1 & I \\ -\tilde{V}_1 & 0 \end{array}\right] \left[\begin{array}{cc} Z & 0 \\ 0 & I \end{array}\right].$$

This establishes the SSE of $\tilde{\Sigma}_1$ and $\tilde{\Sigma}_2$, and consequently, of Σ_1 and Σ_2. $\square$

4.3 System Isomorphisms

Let $\Sigma_i = (T_i, U_i, V_i = H_i T_i, W_i)$ be two admissible systems. The sizes of the matrices are supposed to be $\rho_i \times n_i$, $\rho_i \times m$, $p \times n_i$, $p \times m$, respectively. Define the behaviors

$$B_i = \left\{ \left[\begin{array}{c} x_i \\ u \end{array}\right] \in \mathcal{A}^{n_i + m}, T_i x_i = U_i u \right\}.$$

Definition 27 *An* **admissible system homomorphism** *from Σ_1 to Σ_2 is a mapping*

$$f: \quad B_1 \longrightarrow B_2, \quad \begin{bmatrix} x_1 \\ u \end{bmatrix} \longmapsto \begin{bmatrix} x_2 \\ u \end{bmatrix} := \begin{bmatrix} Q_r x_1 - R_r u \\ u \end{bmatrix} \qquad (4.13)$$

such that

$$V_1 x_1 + W_1 u = V_2 x_2 + W_2 u \quad \text{for all } (x_1, u) := \begin{bmatrix} x_1 \\ u \end{bmatrix} \in B_1, \qquad (4.14)$$

where Q_r and R_r are polynomial matrices of dimensions $n_2 \times n_1$, $n_2 \times m$, respectively.

Lemma 4.3.1. *There exists an admissible system homomorphism from Σ_1 to Σ_2 iff (4.5) holds with polynomial intertwining matrices.*

Proof. The map (4.13) is well-defined iff $(x_2, u) \in B_2$ for all $(x_1, u) \in B_1$. But this is equivalent to

$$T_1 x_1 = U_1 u \quad \Rightarrow \quad T_2 x_2 = T_2(Q_r x_1 - R_r u) = U_2 u$$

or

$$\begin{bmatrix} T_1 & , & -U_1 \end{bmatrix} \begin{bmatrix} x_1 \\ u \end{bmatrix} = 0 \quad \Rightarrow \quad \begin{bmatrix} T_2 Q_r & , & -T_2 R_r - U_2 \end{bmatrix} \begin{bmatrix} x_1 \\ u \end{bmatrix} = 0.$$

This is true iff there exists a polynomial matrix Q_l such that

$$\begin{bmatrix} T_2 Q_r & , & -T_2 R_r - U_2 \end{bmatrix} = Q_l \begin{bmatrix} T_1 & , & -U_1 \end{bmatrix}.$$

This establishes (4.6) and (4.7). Furthermore, from (4.14),

$$T_1 x_1 = U_1 u \quad \Rightarrow \quad V_1 x_1 + W_1 u = V_2(Q_r x_1 - R_r u) + W_2 u.$$

This is equivalent to the existence of a polynomial matrix R_l with

$$\begin{bmatrix} V_1 - V_2 Q_r & , & V_2 R_r + W_1 - W_2 \end{bmatrix} = R_l \begin{bmatrix} T_1 & , & -U_1 \end{bmatrix}$$

hence (4.8) and (4.9). $\qquad \qquad \square$

We will shortly be able to characterize admissible system isomorphisms in terms of strict system equivalence in the Fuhrmann sense. First, we find conditions for an admissible system homomorphism to be injective and surjective. The following lemma should be compared to the corresponding result for one-dimensional systems [32, p. 793].

Lemma 4.3.2. *Let f be an admissible system homomorphism from Σ_1 to Σ_2. Then*

1. f *is injective iff* T_1, Q_r *are zero right co-prime;*
2. f *is surjective iff* $\begin{bmatrix} -Q_l, & T_2 \end{bmatrix}$ *is a minimal left annihilator of*

$$M := \begin{bmatrix} T_1 \\ Q_r \end{bmatrix}.$$

In particular, this implies that Q_l, T_2 *are generalized factor left co-prime and that* $\rho_1 + n_2 = \rho_2 + \operatorname{rank}(M).$
3. *If* f *is bijective, its inverse is again of the form (4.13), i.e.,*

$$f^{-1}: \quad \mathcal{B}_2 \longrightarrow \mathcal{B}_1, \quad \begin{bmatrix} x_2 \\ u \end{bmatrix} \longmapsto \begin{bmatrix} x_1 \\ u \end{bmatrix} := \begin{bmatrix} \tilde{Q}_r x_2 - \tilde{R}_r u \\ u \end{bmatrix}$$

for some polynomial matrices $\tilde{Q}_r$ *and* $\tilde{R}_r$.

Proof.

1. The homomorphism f is injective iff $f(x_1, u) = (Q_r x_1 - R_r u, u) = (0,0)$ for some $(x_1, u) \in \mathcal{B}_1$ implies that $(x_1, u) = (0,0)$. This amounts to

$$\begin{bmatrix} T_1 \\ Q_r \end{bmatrix} x_1 = 0 \quad \Rightarrow \quad x_1 = 0$$

which holds iff T_1, Q_r are zero right co-prime.
2. First note that u is a vector of free variables in $\mathcal{B}_i$, that is, for every $u \in \mathcal{A}^m$, there exists $x_i \in \mathcal{A}^{n_i}$ such that $T_i x_i = U_i u$, or equivalently, $(x_i, u) \in \mathcal{B}_i$. We have

$$\operatorname{im}(f) = \left\{ \begin{bmatrix} x_2 \\ u \end{bmatrix}, \exists x_1 \text{ such that } \begin{bmatrix} T_1 \\ Q_r \end{bmatrix} x_1 = \begin{bmatrix} 0 & U_1 \\ I & R_r \end{bmatrix} \begin{bmatrix} x_2 \\ u \end{bmatrix} \right\}.$$
$$(4.15)$$

Let $\begin{bmatrix} -E, & F \end{bmatrix}$ be a minimal left annihilator of M, then

$$\operatorname{im}(f) = \left\{ \begin{bmatrix} x_2 \\ u \end{bmatrix}, \begin{bmatrix} F, & FR_r - EU_1 \end{bmatrix} \begin{bmatrix} x_2 \\ u \end{bmatrix} = 0 \right\}.$$

Now $\operatorname{im}(f) = \mathcal{B}_2$ iff the row modules of $\begin{bmatrix} F, & FR_r - EU_1 \end{bmatrix}$ and $\begin{bmatrix} T_2, & -U_2 \end{bmatrix}$ coincide. If $\begin{bmatrix} -Q_l, & T_2 \end{bmatrix}$ is a minimal left annihilator of M, we may choose $E = Q_l, F = T_2$, hence $FR_r - EU_1 = T_2 R_r - Q_l U_1 = -U_2$, hence f is surjective. Conversely, suppose that there exists a polynomial matrix Z such that

$$\begin{bmatrix} F, & FR_r - EU_1 \end{bmatrix} = Z \begin{bmatrix} T_2, & -U_2 \end{bmatrix}.$$

From $(E - ZQ_l)T_1 = ET_1 - ZT_2 Q_r = ET_1 - FQ_r = 0$ and the fact that T_1 has full row rank, it follows that $E = ZQ_l$. Together with $F = ZT_2$

from above, this implies that $[\ -Q_l, \quad T_2 \ ]$ is a minimal left annihilator of M.

Recall from Theorem 8 that a minimal left annihilator is generalized factor left prime. If R is a minimal annihilator of M, we have that rank(R)+ rank(M) equals the number of rows of M, which is $\rho_1 + n_2$ in the present situation. Finally, note that rank$[\ -Q_l, \quad T_2 \ ] = $ rank$(T_2) = \rho_2$.

3. If f is injective, there exist polynomial matrices A and B that satisfy the Bézout equation

$$AT_1 + BQ_r = I.$$

If f is surjective, then for any $(x_2, u) \in \mathcal{B}_2$ there exists x_1 such that

$$\begin{bmatrix} T_1 \\ Q_r \end{bmatrix} x_1 = \begin{bmatrix} 0 & U_1 \\ I & R_r \end{bmatrix} \begin{bmatrix} x_2 \\ u \end{bmatrix}.$$

Thus

$$x_1 = Bx_2 + (AU_1 + BR_r)u$$

and the result follows with $\tilde{Q}_r = B$ and $\tilde{R}_r := -(AU_1 + BR_r)$.

$\square$

Theorem 20 *There exists an admissible system isomorphism between Σ_1 and Σ_2 iff Σ_1 and Σ_2 are FSSE.*

Proof. Let f be a bijective admissible system homomorphism from Σ_1 to Σ_2. By Lemma 4.3.1, there exists a relation (4.5) with polynomial intertwining matrices. By Lemma 4.3.2, the injectivity of f implies that T_1, Q_r are zero right co-prime, in particular, rank$(M) = n_1$. Then surjectivity of f implies that $\rho_1 + n_2 = \rho_2 + n_1$. It remains to be shown that Q_l, T_2 are zero left co-prime. But according to Lemma 4.3.3 below, the minimal left annihilators of a zero right prime matrix are zero left prime provided that they have full row rank.

Conversely, let Σ_1 and Σ_2 be FSSE with (4.5) and define f via (4.13). By Lemma 4.3.2, the homomorphism f is injective and the second part of Lemma 4.3.3 implies surjectivity. $\square$

Lemma 4.3.3.

1. *Let T_1, Q_r be zero right co-prime. Then there exist polynomial matrices C and D such that*

$$\begin{bmatrix} T_1 & -C \\ Q_r & D \end{bmatrix}$$

is unimodular. Let A, B, Q_l, and T_2 be defined by

$$\begin{bmatrix} A & B \\ -Q_l & T_2 \end{bmatrix} = \begin{bmatrix} T_1 & -C \\ Q_r & D \end{bmatrix}^{-1} \qquad (4.16)$$

with the obvious partition (i.e., AT_1 a square matrix). Then $\begin{bmatrix} -Q_l, & T_2 \end{bmatrix}$ is a minimal left annihilator (MLA) of

$$M = \begin{bmatrix} T_1 \\ Q_r \end{bmatrix}$$

and any full row rank MLA of M is zero left prime.

2. *Let Q_l, T_1, T_2, Q_r be polynomial matrices of dimensions $\rho_2 \times \rho_1$, $\rho_1 \times n_1$, $\rho_2 \times n_2$, $n_2 \times n_1$ respectively, where $\rho_1 + n_2 = \rho_2 + n_1$. Suppose that $Q_l T_1 = T_2 Q_r$ with Q_l, T_2 zero left co-prime and T_1, Q_r zero right co-prime. Then $\begin{bmatrix} -Q_l, & T_2 \end{bmatrix}$ is a MLA of M.*

Proof.

1. It follows from Lemma 1.3.4 that $\begin{bmatrix} -Q_l, & T_2 \end{bmatrix}$ is a MLA of M. As it can be incorporated row-wise into a unimodular matrix (4.16), it is zero left prime. But two full row rank MLAs of M differ only by a unimodular left factor, hence they are all ZLP.

2. Let $AT_1 + BQ_r = I$ and $Q_l \tilde{C} + T_2 \tilde{D} = I$ be corresponding Bézout equations. Then

$$\begin{bmatrix} A & B \\ -Q_l & T_2 \end{bmatrix} \begin{bmatrix} T_1 & -\tilde{C} \\ Q_r & \tilde{D} \end{bmatrix} = \begin{bmatrix} I_{n_1} & J \\ 0 & I_{\rho_2} \end{bmatrix}$$

where $J = B\tilde{D} - A\tilde{C}$. The matrices on the left are square, and hence unimodular, due to the compatibility constraint $\rho_1 + n_2 = \rho_2 + n_1$. Thus we have

$$\begin{bmatrix} A & B \\ -Q_l & T_2 \end{bmatrix}^{-1} = \begin{bmatrix} T_1 & -\tilde{C} \\ Q_r & \tilde{D} \end{bmatrix} \begin{bmatrix} I & -J \\ 0 & I \end{bmatrix} = \begin{bmatrix} T_1 & -C \\ Q_r & D \end{bmatrix}$$

where $C = T_1 J + \tilde{C}$ and $D = \tilde{D} - Q_r J$, and the result follows from the first part.

□

Finally, we turn to a generalization of a theorem of Pernebo [52, p. 34].

Theorem 21 *Let $\Sigma_1 = (T_1, U_1, V_1 = H_1 T_1, W_1)$ be an admissible system and*

$$\mathcal{B}_1 = \{(x_1, u) \in \mathcal{A}^{n_1} \times \mathcal{A}^m, T_1 x_1 = U_1 u\}.$$

Let Q_r and R_r be two polynomial matrices of appropriate size such that T_1, Q_r are zero right co-prime. Define a mapping

$$f: \quad \mathcal{B}_1 \longrightarrow \mathrm{im}(f), \quad \begin{bmatrix} x_1 \\ u \end{bmatrix} \longmapsto \begin{bmatrix} Q_r x_1 - R_r u \\ u \end{bmatrix}.$$

Then there exists an admissible system $\Sigma_2 = (T_2, U_2, V_2, W_2)$ such that

$$\mathrm{im}(f) = \mathcal{B}_2 = \{(x_2, u) \in \mathcal{A}^{n_2} \times \mathcal{A}^m, \; T_2 x_2 = U_2 u\},$$

and f is a admissible system isomorphism, i.e., Σ_1 and Σ_2 are FSSE.

Proof. As in Lemma 4.3.3, let

$$\begin{bmatrix} A & B \\ -Q_l & T_2 \end{bmatrix} = \begin{bmatrix} T_1 & -C \\ Q_r & D \end{bmatrix}^{-1} \tag{4.17}$$

In particular,

$$Q_l T_1 = T_2 Q_r, \tag{4.18}$$

in fact, we have seen in Lemma 4.3.3 that $\begin{bmatrix} -Q_l, & T_2 \end{bmatrix}$ is even a minimal left annihilator of

$$M = \begin{bmatrix} T_1 \\ Q_r \end{bmatrix}.$$

As in (4.15), we have

$$\mathrm{im}(f) = \left\{ \begin{bmatrix} x_2 \\ u \end{bmatrix}, \quad \exists x_1 \text{ such that } \begin{bmatrix} T_1 \\ Q_r \end{bmatrix} x_1 = \begin{bmatrix} 0 & U_1 \\ I & R_r \end{bmatrix} \begin{bmatrix} x_2 \\ u \end{bmatrix} \right\}.$$

As $\begin{bmatrix} -Q_l, & T_2 \end{bmatrix}$ is a minimal left annihilator of M, we obtain

$$\mathrm{im}(f) = \left\{ \begin{bmatrix} x_2 \\ u \end{bmatrix}, \quad \begin{bmatrix} T_2 & , & -Q_l U_1 + T_2 R_r \end{bmatrix} \begin{bmatrix} x_2 \\ u \end{bmatrix} = 0 \right\}.$$

Define

$$U_2 := Q_l U_1 - T_2 R_r, \tag{4.19}$$

$R_l := V_1 A$, $V_2 := V_1 B$ and

$$W_2 := R_l U_1 + V_2 R_r + W_1. \tag{4.20}$$

Then according to (4.17),

$$R_l T_1 + V_2 Q_r = V_1 (A T_1 + B Q_r) = V_1. \tag{4.21}$$

Equations (4.18) to (4.21) are precisely the constituent equations of (4.5) with polynomial intertwining matrices. It remains to be shown that $\Sigma_2 = (T_2, U_2, V_2, W_2)$ is admissible, then f is an admissible system homomorphism.

As injectivity and surjectivity of f follow directly from the assumptions, this yields the desired result.

Considering (4.17) and taking Schur complements, it follows that T_2 has full row rank, similarly as in the proof of Theorem 19. Finally, from (4.17), we have $T_1 B = C T_2$, and hence

$$V_2 = V_1 B = H_1 T_1 B = H_1 C T_2 = H_2 T_2$$

where $H_2 := H_1 C$. $\qquad\qquad\qquad\qquad\qquad\qquad\qquad\qquad\qquad\qquad$ □

Appendix

Schur complements. Let

$$A = \begin{bmatrix} A_{11} & A_{12} \\ A_{21} & A_{22} \end{bmatrix} = \begin{bmatrix} B_{11} & B_{12} \\ B_{21} & B_{22} \end{bmatrix}^{-1}$$

where A_{ii} and B_{ii} are square. Then B_{11} is invertible iff A_{22} is invertible, indeed, we have

$$\det(A) \det(B_{11}) = \det(A_{22}).$$

Relation between FSSE and RSSE. Two admissible systems with $\rho_1 + n_2 = \rho_2 + n_1$ are FSSE iff a trivial expansion of one system is RSSE to a trivial expansion of the other. The proof is given along the lines of that in [57].

"if": Suppose that

$$\begin{bmatrix} Q_{11} & Q_{12} & 0 \\ Q_{13} & Q_{14} & 0 \\ R_{11} & R_{12} & I \end{bmatrix} \begin{bmatrix} I & 0 & 0 \\ 0 & T_1 & U_1 \\ 0 & -V_1 & W_1 \end{bmatrix} =$$

$$= \begin{bmatrix} I & 0 & 0 \\ 0 & T_2 & U_2 \\ 0 & -V_2 & W_2 \end{bmatrix} \begin{bmatrix} Q_{21} & Q_{22} & R_{21} \\ Q_{23} & Q_{24} & R_{22} \\ 0 & 0 & I \end{bmatrix}$$

with unimodular matrices

$$Q_l = \begin{bmatrix} Q_{11} & Q_{12} \\ Q_{13} & Q_{14} \end{bmatrix} \quad \text{and} \quad Q_r = \begin{bmatrix} Q_{21} & Q_{22} \\ Q_{23} & Q_{24} \end{bmatrix}$$

Then

$$\begin{bmatrix} Q_{14} & 0 \\ R_{12} & I \end{bmatrix} \begin{bmatrix} T_1 & U_1 \\ -V_1 & W_1 \end{bmatrix} = \begin{bmatrix} T_2 & U_2 \\ -V_2 & W_2 \end{bmatrix} \begin{bmatrix} Q_{24} & R_{22} \\ 0 & I \end{bmatrix}$$

and

$$\begin{bmatrix} Q_{11} & Q_{12} \\ Q_{13} & Q_{14} \end{bmatrix} \begin{bmatrix} I & 0 \\ 0 & T_1 \end{bmatrix} = \begin{bmatrix} I & 0 \\ 0 & T_2 \end{bmatrix} \begin{bmatrix} Q_{21} & Q_{22} \\ Q_{23} & Q_{24} \end{bmatrix}$$

As Q is unimodular, $Q_{13} = T_2 Q_{23}$ and Q_{14} are zero left co-prime, hence T_2 and Q_{14} are zero left co-prime. Similarly, it follows that T_1 and Q_{24} are zero right co-prime.

"only if": Suppose that

$$\begin{bmatrix} Q_l & 0 \\ R_l & I \end{bmatrix} \begin{bmatrix} T_1 & U_1 \\ -V_1 & W_1 \end{bmatrix} = \begin{bmatrix} T_2 & U_2 \\ -V_2 & W_2 \end{bmatrix} \begin{bmatrix} Q_r & R_r \\ 0 & I \end{bmatrix}$$

where Q_l, T_2 are zero left co-prime and T_1, Q_r are zero right co-prime. Let the corresponding Bézout relations be $Q_l A + T_2 B = I$ and $C T_1 + D Q_r = I$. Then

$$\begin{bmatrix} -D & C \\ T_2 & Q_l \end{bmatrix} \begin{bmatrix} -Q_r & B \\ T_1 & A \end{bmatrix} = \begin{bmatrix} I & CA - DB \\ 0 & I \end{bmatrix}$$

where the matrices on the left hand side are square, and hence unimodular, due to the compatibility constraint $\rho_1 + n_2 = \rho_2 + n_1$. Now

$$\begin{bmatrix} -D & C & 0 \\ T_2 & Q_l & 0 \\ -V_2 & R_l & I \end{bmatrix} \begin{bmatrix} I_{n_2} & 0 & 0 \\ 0 & T_1 & U_1 \\ 0 & -V_1 & W_1 \end{bmatrix} =$$

$$= \begin{bmatrix} I_{n_1} & 0 & 0 \\ 0 & T_2 & U_2 \\ 0 & -V_2 & W_2 \end{bmatrix} \begin{bmatrix} -D & CT_1 & CU_1 \\ I & Q_r & R_r \\ 0 & 0 & I \end{bmatrix}$$

It remains to be shown that

$$\begin{bmatrix} -D & CT_1 \\ I & Q_r \end{bmatrix}$$

is unimodular. But this follows from

$$\begin{bmatrix} -D & CT_1 \\ I & Q_r \end{bmatrix} = \begin{bmatrix} -D & I - DQ_r \\ I & Q_r \end{bmatrix} = \begin{bmatrix} -D & I \\ I & 0 \end{bmatrix} \begin{bmatrix} I & Q_r \\ 0 & I \end{bmatrix}.$$

5. First Order Representations of Multidimensional Systems

Several first order representations for multidimensional systems have been proposed in the literature, mostly in the two-dimensional case. The best-known model is probably the one introduced by Roesser [62]:

$$\begin{bmatrix} x^h(i+1,j) \\ x^v(i,j+1) \end{bmatrix} = \begin{bmatrix} A_1 & A_2 \\ A_3 & A_4 \end{bmatrix} \begin{bmatrix} x^h(i,j) \\ x^v(i,j) \end{bmatrix} + \begin{bmatrix} B_1 \\ B_2 \end{bmatrix} u(i,j)$$

$$y(i,j) = \begin{bmatrix} C_1 & C_2 \end{bmatrix} \begin{bmatrix} x^h(i,j) \\ x^v(i,j) \end{bmatrix} + Du(i,j).$$

Here, x^h denotes the horizontal, and x^v the vertical local state vector. Input and output are denoted by u and y, respectively. All signals are defined on the two-dimensional lattice $\mathbb{N}^2$. Another famous model is the Fornasini–Marchesini model [22]

$$\begin{aligned} x(i+1,j+1) &= A_1 x(i+1,j) + A_2 x(i,j+1) + \\ & \quad B_1 u(i+1,j) + B_2 u(i,j+1) \\ y(i,j) &= Cx(i,j) + Du(i,j), \end{aligned}$$

which is usually defined over $\mathbb{Z}^2$. These models and variants of them have been studied by several authors; for a survey see [33]. A serious drawback of these local state-space models is that they require quarter-plane causality, *i.e.*, the corresponding input-output relation (assuming zero "initial" conditions) can be formally written as

$$y(i,j) = \sum_{k=-\infty}^{i} \sum_{l=-\infty}^{j} h(i-k,j-l)u(k,l),$$

where h is a suitably defined impulse response. One way to get rid of the causality requirement is to admit singular systems: As is well-known for one-dimensional systems, only proper rational matrices admit a representation $C(sI - A)^{-1}B + D$, but every rational matrix can be written in the form $C(sE - A)^{-1}B + D$. An analogous fact holds for singular multidimensional Roesser models, which have been studied by Kaczorek [34] and Gałkowski [26].

Here, we adopt another approach to first order representations, which is based on Willems' "state–space behaviors" [81, 58] in "dual pencil representation" [36]. A vector w of manifest system variables (e.g., "inputs and outputs") and a vector x of latent system variables (e.g., generalized "states" that arise in modeling or with the reduction to first order) are supposed to be related via a first order dynamic equation and a static equation. In the 1D case, this amounts to the model

$$
\begin{aligned}
x(i+1) &= Ax(i) + Bw(i) & (5.1)\\
0 &= Cx(i) + Dw(i). & (5.2)
\end{aligned}
$$

Note that even if we take $w = \begin{bmatrix} u^T & y^T \end{bmatrix}^T$, this is more general than the usual input-state-output model, as the static equations need not be solvable for y. The continuous counterpart of these "output–nulling representations" has been studied by Weiland [78].

In the multidimensional setting, a generalization of state-space behaviors in dual pencil representation has been introduced by Miri and Aplevich [45]. Our model can be embedded into theirs, but the model class we consider is interesting by itself as it is large enough to comprise all "autoregressive" models, *i.e.*, systems given by linear partial difference equations with constant coefficients.

Moreover, the dynamic equation (5.1) of an ON representation is regular in the sense that it is explicit for the updated "state" $x(i+1)$, and this advantage carries over to multidimensional systems. This makes it possible to evade several difficulties that are inherent in singular (*i.e.*, implicit) models.

5.1 Preliminaries and Notation

We will consider the discrete signal domain $\mathbb{Z}^r$. For finite subsets $T \subset \mathbb{Z}^r$, let $\vee(T) \in \mathbb{Z}^r$ denote the **component-wise (cw-) supremum** of T, defined by

$$
\vee(T)_i = \max \pi_i(T) \quad \text{for } 1 \leq i \leq r,
$$

where $\pi_i : \mathbb{Z}^r \to \mathbb{Z}, (t_1, \ldots, t_r) \mapsto t_i$, denotes the projection onto the i-th component. Similarly, the **cw-infimum** is defined by

$$
\wedge(T)_i = \min \pi_i(T) \quad \text{for } 1 \leq i \leq r.
$$

The action of the **shift operators** $\sigma^t = \sigma_1^{t_1} \cdots \sigma_r^{t_r}$, where $t = (t_1, \ldots, t_r) \in \mathbb{Z}^r$, is defined by

$$
(\sigma^t w)(t') = w(t + t') \quad \text{for all } t' \in \mathbb{Z}^r \tag{5.3}
$$

and for all sequences $w : \mathbb{Z}^r \to \mathbb{F}^q$, where $\mathbb{F}$ is an arbitrary field. A linear shift-invariant difference **behavior** in kernel representation is given by

$$\mathcal{B} = \{w : \mathbb{Z}^r \to \mathbb{F}^q, \ R(\sigma, \sigma^{-1})w = 0\}, \quad R \in \mathbb{F}[s, s^{-1}]^{g \times q}. \qquad (5.4)$$

The Laurent polynomial matrix R is called a **kernel representation** of $\mathcal{B}$. We may write

$$R = \sum_{t \in \mathrm{supp}(R)} R_t s^t,$$

where $\mathrm{supp}(R) \subset \mathbb{Z}^r$ denotes the finite set of multi-indices t which appear in the sum above with a non-zero coefficient matrix $R_t \in \mathbb{F}^{g \times q}$.

A behavior has infinitely many different kernel representations, indeed, let $\mathcal{B}$ be represented by R according to (5.4). Then a matrix R' with g' rows represents the same behavior iff its rows generate the same module over $\mathbb{F}[s, s^{-1}]$, that is, iff

$$\mathbb{F}[s, s^{-1}]^{1 \times g} R = \mathbb{F}[s, s^{-1}]^{1 \times g'} R'.$$

This is true iff there exist Laurent polynomial matrices X and Y such that $R = XR'$ and $R' = YR$. In particular, all matrices in the class

$$\{UR; \ U \text{ unimodular}\}$$

represent the same behavior. Here, the term **unimodular** refers to invertibility in $\mathbb{F}[s, s^{-1}]^{g \times g}$, that is, a square Laurent polynomial matrix U is unimodular iff

$$\det(U) = fs^t \quad \text{for some } f \in \mathbb{F} \setminus \{0\}, \ t \in \mathbb{Z}^r.$$

For instance, we can pre-multiply R by

$$\mathrm{diag}(s^{-k_1}, \dots, s^{-k_g}), \qquad (5.5)$$

where each k_i denotes the cw-supremum of the support of the i-th row of R,

$$k_i = \vee(\mathrm{supp}(R_{i-})). \qquad (5.6)$$

Thus, the representing matrix R can be chosen such that it contains only non-positive powers of the indeterminates $s_1, \dots, s_r$, that is, without loss of generality, we can assume that

$$R = R(s^{-1}) \in \mathbb{F}[s^{-1}]^{g \times q} = \mathbb{F}[s_1^{-1}, \dots, s_r^{-1}]^{g \times q}.$$

Similarly, the exponents of the entries of R can be shifted into any of the 2^r hyper-quadrants of $\mathbb{Z}^r$ defined by a subset $I \subseteq \{1, \dots, r\}$ and

$$t_i \geq 0 \text{ for } i \in I \quad \text{and} \quad t_i \leq 0 \text{ for } i \notin I.$$

Chapter 6 shows that $R(s^{-1}) \in \mathbb{F}[s^{-1}]^{g \times q}$ can be written in terms of a linear fractional transformation (LFT)

$$\begin{aligned} R(s^{-1}) &= \mathcal{F}(\Delta(s^{-1}), M) \\ &= C\Delta(s^{-1})(I - A\Delta(s^{-1}))^{-1}B + D \\ &= C(\Delta(s) - A)^{-1}B + D, \end{aligned}$$

where for some suitable integers $n_1, \dots, n_r \geq 0$ and $n := \sum_{i=1}^{r} n_i$,

$$\Delta(s) = \text{diag}(s_1 I_{n_1}, \dots, s_r I_{n_r}), \quad \Delta(s^{-1}) = \Delta(s)^{-1}, \quad \text{and}$$

$$M = \begin{bmatrix} A & B \\ C & D \end{bmatrix} \in \mathbb{F}^{(n+g) \times (n+q)}. \tag{5.7}$$

Since R is given in terms of $n = (n_1, \dots, n_r)$ and A, B, C, D, the following notation will be convenient:

$$R = \mathcal{F}(n; A, B, C, D).$$

Moreover, the LFT construction of Chapter 6 will yield that the matrix A is such that

$$\det(I - A\Delta(s^{-1})) = 1.$$

This implies that $\det(\Delta(s) - A) = \det(I - A\Delta(s^{-1})) \det(\Delta(s)) = s_1^{n_1} \cdots s_r^{n_r}$, i.e., $\Delta(s) - A$ is invertible over the Laurent polynomial ring.

Theorem 22 *Let $r \geq 1$ and $n = (n_1, \dots, n_r) \in \mathbb{N}^r$ be given. Define $n := \sum_{i=1}^{r} n_i$ and $\Delta(s) = \text{diag}(s_1 I_{n_1}, \dots, s_r I_{n_r})$. Let $A \in \mathbb{F}^{n \times n}$ be such that $\Delta(s) - A$ is invertible over the Laurent polynomial ring, i.e.,*

$$\det(\Delta(s) - A) = s^n := s_1^{n_1} \cdots s_r^{n_r}.$$

Then, with the partition corresponding to n, that is

$$A = \begin{bmatrix} A_{11} & \cdots & A_{1r} \\ \vdots & & \vdots \\ A_{r1} & \cdots & A_{rr} \end{bmatrix}, \quad A_{ij} \in \mathbb{F}^{n_i \times n_j},$$

the matrices A_{ii} are nilpotent for $1 \leq i \leq r$.

Proof. The proof is by induction on r. For $r = 1$, $\det(sI_n - A_{11}) = s^n$ implies that A_{11} is nilpotent. Suppose that the statement is true for $r - 1$. Let $\Delta(s') = \text{diag}(s_1 I_{n_1}, \dots, s_{r-1} I_{n_{r-1}})$. With the corresponding partition of A,

$$\Delta(s) - A = \begin{bmatrix} \Delta(s') - A' & -A_{(1,2)} \\ -A_{(2,1)} & s_r I_{n_r} - A_{rr} \end{bmatrix}$$

and hence, assuming that $\det(\Delta(s) - A) = s^n$,

$$s_1^{n_1} \cdots s_r^{n_r} = \det(\Delta(s') - A')\det(s_r I_{n_r} - A_{rr} - A_{(2,1)}(\Delta(s') - A')^{-1}A_{(1,2)}).$$
$$(5.8)$$

Viewing (5.8) as an identity over $\mathbb{F}(s_1, \ldots, s_{r-1})[s_r]$, we may compare the coefficients of $s_r^{n_r}$ on both sides of the equation to get

$$s_1^{n_1} \cdots s_{r-1}^{n_{r-1}} = \det(\Delta(s') - A').$$

The inductive hypothesis implies that A_{ii} are nilpotent for $1 \leq i \leq r-1$. Similarly,

$$s_1^{n_1} \cdots s_r^{n_r} = \det(s_r I_{n_r} - A_{rr})\det(\Delta(s') - A' - A_{(1,2)}(s_r I_{n_r} - A_{rr})^{-1}A_{(2,1)})$$

implies that $\det(s_r I_{n_r} - A_{rr}) = s_r^{n_r}$, hence A_{rr} is nilpotent.   □

5.2 Output–nulling Representations

A linear fractional representation of $R(s^{-1}) \in \mathbb{F}[s^{-1}]^{g \times q}$,

$$R(s^{-1}) = C(\Delta(s) - A)^{-1}B + D$$

gives rise to first order representations of

$$\mathcal{B} = \{w : \mathbb{Z}^r \to \mathbb{F}^q, \ R(\sigma^{-1})w = 0\}$$

in the following way: For a trajectory $w \in \mathcal{B}$, let $x : \mathbb{Z}^r \to \mathbb{F}^n$ be a solution of

$$\Delta(\sigma)x = Ax + Bw. \tag{5.9}$$

With a partition of x that corresponds to $n = (n_1, \ldots, n_r)$, that is, $x = (x_1, \ldots, x_r)^T$ with x_i having n_i components, x solves the following system of first order partial difference equations:

$$
\begin{bmatrix}
x_1(t_1 + 1, t_2, \ldots, t_r) \\
\vdots \\
x_i(t_1, \ldots, t_i + 1, \ldots, t_r) \\
\vdots \\
x_r(t_1, \ldots, t_{r-1}, t_r + 1)
\end{bmatrix}
= A
\begin{bmatrix}
x_1(t_1, \ldots, t_r) \\
\vdots \\
x_i(t_1, \ldots, t_r) \\
\vdots \\
x_r(t_1, \ldots, t_r)
\end{bmatrix}
+ Bw(t_1, \ldots, t_r)
$$

for all $t = (t_1, \ldots, t_r) \in \mathbb{Z}^r$. Then

$$w \in \mathcal{B} \quad \Leftrightarrow \quad \exists x : \mathbb{Z}^r \to \mathbb{F}^n \text{ with } \begin{cases} \Delta(\sigma)x &= Ax + Bw \\ 0 &= Cx + Dw. \end{cases} \tag{5.10}$$

Using that $(\Delta(s) - A)^{-1}$ is a Laurent polynomial (rather than only a rational) matrix, one can rewrite (5.9) as

$$x = (\Delta(\sigma) - A)^{-1} Bw. \tag{5.11}$$

Thus $R(\sigma^{-1})w = Cx + Dw$, which proves the equivalence. Note that by (5.11), the solution x of (5.9) is uniquely determined by w, and that (5.9) constitutes a **finite-memory system** [20].

The right hand side of (5.10) provides an **output-nulling (ON) representation** of $\mathcal{B}$ according to the subsequent definition. The following notation will be used: For $n \in \mathbb{Z}^r$, let $|n| := \sum_{i=1}^{r} |n_i|$. Define a set of matrix quadruples (A, B, C, D) of appropriate dimensions (according to (5.7)) by

$$\mathbb{F}(n, g, q) := \mathbb{F}^{n \times n} \times \mathbb{F}^{n \times q} \times \mathbb{F}^{g \times n} \times \mathbb{F}^{g \times q}.$$

Definition 28 *Let* $(A, B, C, D) \in \mathbb{F}(n, g, q)$ *be given and let* $n = (n_1, \ldots, n_r)$ *in* $\mathbb{N}^r$ *be such that* $|n| = n$. *Define* $\Delta(s) = \mathrm{diag}(s_1 I_{n_1}, \ldots, s_r I_{n_r})$. *The equations*

$$\begin{aligned} \Delta(\sigma)x &= Ax + Bw \tag{5.12} \\ 0 &= Cx + Dw \tag{5.13} \end{aligned}$$

are called an **output-nulling (ON) representation** *of* $\mathcal{B}$ *if*

$$\mathcal{B} = \{w : \mathbb{Z}^r \to \mathbb{F}^q, \exists x : \mathbb{Z}^r \to \mathbb{F}^n \text{ such that } (5.12), (5.13) \text{ hold}\}.$$

For simplicity, the data quintuple $(n; A, B, C, D)$ *itself is referred to as an ON representation. The* **full behavior** *associated with the ON representation is*

$$\mathcal{B}_{ON}(n; A, B, C, D) = \{(x, w) : \mathbb{Z}^r \to \mathbb{F}^n \times \mathbb{F}^q, x, w \text{ satisfy } (5.12), (5.13)\}.$$

The notation $\mathcal{B} = \pi_w \mathcal{B}_{ON}(n; A, B, C, D)$ *will be used, where* π_w *denotes the projection onto the second component.*

If $(n; A, B, C, D)$ is an ON representation of $\mathcal{B}$, then so is

$$(n; \ T^{-1}(A + LC)T, \ T^{-1}(B + LD), \ PCT, \ PD), \tag{5.14}$$

there $T = \mathrm{diag}(T_1, \ldots, T_r)$, $T_i \in \mathbb{F}^{n_i \times n_i}$ and $P \in \mathbb{F}^{g \times g}$ are non-singular, and L is an arbitrary $n \times g$ matrix over $\mathbb{F}$.

Next, we introduce a notion of minimality for ON representations. As usual in multidimensional systems theory, it is hard to give necessary and sufficient conditions for minimality (in the "minimum number of equations" sense, which must not be confused with minimality in the transfer class). We have already encountered this problem when discussing kernel representations of a behavior $\mathcal{B} = \ker(R)$ with a minimum number of equations (rows of R). Recall that the behavior $\mathcal{B}$ and the module generated by the rows of its representation matrix R are equivalent data. Thus, the question of finding $g_{ker}(\mathcal{B})$, the smallest possible number of autoregressive equations defining $\mathcal{B}$, is equivalent to finding the minimal number of generators for the module $\mathbb{F}[s, s^{-1}]^{1 \times g} R$. The problem of efficient generation of modules over (Laurent) polynomial rings has been addressed by many authors (e.g., Mandal [44] or Kumar [37], who proved a famous conjecture of Eisenbud and Evans on an upper bound for $g_{ker}(\mathcal{B})$).

On the other hand, it is clear that $g_{ker}(\mathcal{B}) \geq \text{rank}(R)$, where $\text{rank}(R)$ is an invariant of the behavior $\mathcal{B}$ (i.e., independent of the particular choice of the representation), usually referred to as the **output-dimension** of $\mathcal{B}$ [48]. In the case $r = 1$, there is no loss of generality in assuming that $g_{ker}(\mathcal{B}) = \text{rank}(R)$, and for $r = 2$, we have the desired result at least for behaviors that are minimal in their transfer class, i.e., controllable. This is not true for $r > 2$ however, as can be seen from the example in Section 1.2.2. Concerning the minimality problem (in the "minimum number of equations" sense) with multidimensional systems, one is usually content with easily testable necessary conditions that lead to constructive reduction techniques.

An ON representation is said to be **minimal** iff the pair $(n, g) \in \mathbb{N}^2$, where n denotes the dimension of the state space, and g denotes the number of output-nulling equations, is minimal with respect to the component-wise order on $\mathbb{N}^2$. In other words:

Definition 29 *An ON representation* $(n; A, B, C, D)$ *of* $\mathcal{B}$ *is called* **minimal** *if*

$$\left. \begin{array}{l} \mathcal{B} = \pi_w \mathcal{B}_{ON}(n', A', B', C', D') \text{ for some } n' \in \mathbb{N}^r, \\[2mm] (A', B', C', D') \in \mathbb{F}(n', g', q), \quad n' = |n'| \end{array} \right\} \quad (5.15)$$

implies that $n < n'$ *or* $g < g'$ *or* $(n, g) = (n', g')$. *If the implication of (5.15) is even* $n \leq n'$ *and* $g \leq g'$, *we say that the ON representation has* **least order**.

It is clear that every behavior possesses minimal ON representations, but a behavior does not necessarily have a least order representation, i.e., one in which n and g both attain their respective minimum. Take for instance, $\mathcal{B} = \mathbb{F}^{\mathbb{Z}}$, which can be represented by the trivial equation $0w = 0$ ($n = 0$, $g = 1$), but also by $\exists x : \sigma_1 x = w$ ($n = 1$, $g = 0$).

When comparing two ON representation with the same number of output-nulling equations, **state-minimality** becomes the crucial concept:

Definition 30 *An ON representation* $(n; A, B, C, D)$ *of* $\mathcal{B}$ *is called* **state-minimal** *if (5.15) implies* $n \leq n'$. *We call* $n = |n|$ *the* **dimension** *of the ON representation.*

We first give necessary conditions for state-minimality of an ON representation. In 1D systems theory, an ON representation over $\mathbb{Z}$ is state-minimal iff (C, A) is observable,

$$\text{im}\begin{bmatrix} I \\ 0 \end{bmatrix} \subseteq \text{im}\begin{bmatrix} A & B \\ C & D \end{bmatrix},$$

and $\text{im}(C) \subseteq \text{im}(D)$ [81, p. 270]. Since we may assume, without loss of generality, that $[C, D]$ has full row rank, these conditions are equivalent to: (C, A) is observable, and the matrices

$$\begin{bmatrix} A & B \\ C & D \end{bmatrix}$$

and D both have full row rank.

Lemma 5.2.1. *Let $(n; A, B, C, D)$ be a state-minimal ON representation. Without loss of generality, assume that $[C, D]$ has full row rank. With the partition induced by n, we write*

$$A = \begin{bmatrix} A_{11} & \cdots & A_{1r} \\ \vdots & & \vdots \\ A_{r1} & \cdots & A_{rr} \end{bmatrix}, \quad B = \begin{bmatrix} B_1 \\ \vdots \\ B_r \end{bmatrix}, \quad C = [\, C_1 \ \cdots \ C_r \,].$$

Then we have, for $1 \le i \le r$,

1. $\begin{bmatrix} A_{i1} & \cdots & A_{ir} & B_i \\ C_1 & \cdots & C_r & D \end{bmatrix}$ *has full row rank;*

2. $\left([\, A_{1i}^T \ \cdots \ A_{i-1,i}^T \ A_{i+1,i}^T \ \cdots \ A_{ri}^T \ C_i^T \,]^T, A_{ii} \right)$ *is observable.*

Proof. Without loss of generality, let $i = 1$. Suppose that there exist a row vector $[\xi, \eta] \ne 0$ with

$$[\, \xi \ \ \eta \,]\begin{bmatrix} A_{11} & \cdots & A_{1r} & B_1 \\ C_1 & \cdots & C_r & D \end{bmatrix} = 0.$$

If we pre-multiply the system equations

$$\begin{aligned} \sigma_1 x_1 &= A_{11}x_1 + \ldots + A_{1r}x_r + B_1 w \\ 0 &= C_1 x_1 + \ldots + C_r x_r + Dw \end{aligned}$$

by $[\xi, \eta]$, we obtain that $\xi\sigma_1 x_1 = 0$. Then $\xi x_1 = 0$. Since $[C, D]$ is assumed to have full row rank, $\xi \ne 0$. Thus $x_1 \in V := \{v \in \mathbb{F}^{n_1}, \xi v = 0\}$, and $d := \dim V < n_1$. Let $\{e_1, \ldots, e_d\}$ be a basis of V, then there is an isomorphism

$$\mathbb{F}^d \to V, \quad \lambda_1 \mapsto E\lambda_1, \quad E = [\, e_1 \ \cdots \ e_d \,] \in \mathbb{F}^{n_1 \times d}.$$

Any $x_1 \in V$ has the form $x_1 = E\lambda_1$, and E, being a full column rank matrix, possesses a left inverse $E^{\#}$, that is, $E^{\#}E = I$. Then

$$\sigma_1\lambda_1 = E^\# A_{11}E\lambda_1 + E^\# A_{12}x_2 + \ldots + E^\# A_{1r}x_r + E^\# B_1 w$$
$$\sigma_2 x_2 = A_{21}E\lambda_1 + A_{22}x_2 + \ldots + A_{2r}x_r + B_2 w$$
$$\vdots$$
$$\sigma_r x_r = A_{r1}E\lambda_1 + A_{r2}x_2 + \ldots + A_{rr}x_r + B_r w$$
$$0 = C_1 E\lambda_1 + C_2 x_2 + \ldots + C_r x_r + D_w$$

represents the same behavior as the original ON representation, but has a strictly smaller state dimension. The second statement follows from a Kalman observability decomposition. □

The condition $\mathrm{im}(C) \subseteq \mathrm{im}(D)$ fails to generalize to the multidimensional setting. As we may assume $\begin{bmatrix} C & D \end{bmatrix}$ to be a full row rank matrix, the condition $\mathrm{im}(C) \subseteq \mathrm{im}(D)$ amounts to D being a full row rank matrix. Unlike the 1D situation, this assumption is restrictive for multidimensional systems; see the definition of proper ON representations below. In fact, there even are least order ON representations in which $D = 0$.

Example 5.2.1. Consider

$$\mathcal{B} = \{w : \mathbb{Z}^2 \to \mathbb{R}, \ (\sigma_1 + \sigma_2 - 1)w = 0\}.$$

An equivalent kernel representation is $R = \sigma_1^{-1}\sigma_2^{-1} - \sigma_1^{-1} - \sigma_2^{-1}$. Constructing an LFT representation according to Chapter 6 and reducing it according to Section 6.2 yields the ON representation

$$\begin{bmatrix} \sigma_1 & 0 \\ 0 & \sigma_2 \end{bmatrix}\begin{bmatrix} x_1 \\ x_2 \end{bmatrix} = \begin{bmatrix} 0 & 0 \\ -1 & 0 \end{bmatrix}\begin{bmatrix} x_1 \\ x_2 \end{bmatrix} + \begin{bmatrix} 1 \\ 1 \end{bmatrix} w$$

$$0 = \begin{bmatrix} 1 & 1 \end{bmatrix}\begin{bmatrix} x_1 \\ x_2 \end{bmatrix}.$$

This representation of dimension two is state-minimal (any ON representation of dimension one depends either on σ_1 or on σ_2 alone, and thus is certainly incapable of representing $\mathcal{B}$) and even a least order representation, as there is one output-nulling equation (the only behavior that possesses ON representations without output-nulling equations is the whole of $\mathbb{R}^{\mathbb{Z}^2}$).

Definition 31 *An ON representation is called* **observable** *if knowledge of the external signal w yields knowledge of the internal signal x, i.e., iff $w = 0$ implies $x = 0$.*

Lemma 5.2.2. *Let $(n; A, B, C, D)$ be an ON representation. Define the* **Hautus matrix**

$$H := \begin{bmatrix} \Delta(s) - A \\ C \end{bmatrix} \in \mathbb{F}[s, s^{-1}]^{(n+g) \times n}$$

The $n \times n$ minors (sub-determinants) of H are denoted by $h_1, \ldots, h_k$. The ON representation is observable iff the following equivalent conditions are satisfied:

1. $\mathcal{B}_{unobs} := \{x : \mathbb{Z}^r \to \mathbb{F}^n, \ H(\sigma)x = 0\} = \{0\}$;
2. H *is zero right prime, i.e., there exists a Laurent polynomial matrix L such that $LH = I_n$;*
3. *There exist Laurent polynomials $l_1, \ldots, l_k$ such that $\sum_{i=1}^{k} l_i h_i = 1$.*

The proof is analogous to the known case over $\mathbb{F}[s]$, and can be found in [24].

Note that observability is neither sufficient nor necessary for state-minimality. The ON representations constructed in Chapter 6 are always observable due to (5.11), but they are usually *not* state-minimal. On the other hand, an example below shows that even least order ON representations need not be observable.

For discussing minimality with respect to the number of ON equations, we need the following prerequisite: We have seen how to derive an ON representation with g ON equations from a given kernel representation with g autoregressive equations. Conversely, let

$$\mathcal{B} = \left\{ w : \mathbb{Z}^r \to \mathbb{F}^q, \ \exists x : \mathbb{Z}^r \to \mathbb{F}^n : \begin{bmatrix} \Delta(\sigma) - A \\ C \end{bmatrix} x = \begin{bmatrix} B \\ -D \end{bmatrix} w \right\}$$

be given. In order to eliminate the latent variables x, we have to compute a minimal left annihilator $X = \begin{bmatrix} X_1 & X_2 \end{bmatrix}$ of

$$H = \begin{bmatrix} \Delta(s) - A \\ C \end{bmatrix}$$

(over $\mathbb{F}[s, s^{-1}]$) to obtain, by means of the fundamental principle, the kernel representation

$$\mathcal{B} = \{w : \mathbb{Z}^r \to \mathbb{F}^q, \ (X_1 B - X_2 D)(\sigma, \sigma^{-1})w = 0\}. \tag{5.16}$$

Observe that X, being a MLA of a full column rank $(n + g) \times n$ matrix, must have rank g. In particular, it has at least g rows. The case that X can be chosen with precisely g rows can be characterized as follows.

Lemma 5.2.3. *Let $(n; A, B, C, D)$ be an ON representation of $\mathcal{B} \neq (\mathbb{F}^q)^{\mathbb{Z}^r}$. The matrix H as defined above possesses a MLA of full row rank iff the projective dimension (pd) of*

$$\mathrm{cok}(H) = \mathbb{F}[s, s^{-1}]^{1 \times n} / \mathbb{F}[s, s^{-1}]^{1 \times (n+g)} H$$

does not exceed two. In particular, this is true for any ON representation of dimension $r \leq 2$, and for observable ON representations of arbitrary dimension r. This implies that

$$g_o(\mathcal{B}) = g_{ker}(\mathcal{B}),$$

where $g_o(\mathcal{B})$ denotes the minimal number of ON equations in any observable ON representation of $\mathcal{B}$, and $g_{ker}(\mathcal{B})$ is as discussed above.

Proof. Consider the exact sequence

$$0 \to \ker(H) \hookrightarrow \mathbb{F}[s, s^{-1}]^{1 \times (n+g)} \xrightarrow{H} \mathbb{F}[s, s^{-1}]^{1 \times n} \to \mathrm{cok}(H) \to 0.$$

The module $\ker(H)$ is free iff $\mathrm{pd}(\mathrm{cok}(H)) \leq 2$. The case $\ker(H) = 0$ corresponds to $\mathcal{B} = (\mathbb{F}^q)^{\mathbb{Z}^r}$. This being excluded, $\ker(H)$ can be spanned by linearly independent row vectors.

If $r \leq 2$, we get $\mathrm{pd}(\mathrm{cok}(H)) \leq 2$ from Hilbert's syzygy theorem. If H is zero right prime, then $\mathrm{cok}(H) = 0$, and hence its projective dimension is zero. Finally, as kernel representations yield observable ON representations with the same number of equations,

$$g_{\mathrm{ker}}(\mathcal{B}) \geq g_o(\mathcal{B}).$$

Conversely, from an observable ON representation with g ON equations, we compute a full row rank MLA of the associated Hautus matrix and obtain a kernel representation (5.16) with g autoregressive equations. Hence $g_{\mathrm{ker}}(\mathcal{B}) \leq g_o(\mathcal{B})$. □

The first part of the above lemma should be compared with Theorem 12, where a similar argument has been used. In the case of a non-observable ON representation, we have the following inequality (combine [18, p. 455] with [47, p. 59/60])

$$\mathrm{pd}(\mathrm{cok}(H)) \geq \mathrm{codim}(\mathcal{I}(H)),$$

where $\mathcal{I}(H)$ denotes the ideal generated by the $n \times n$ minors of H. Non-observability amounts to $\mathrm{codim}(\mathcal{I}(H)) < \infty$.

Example 5.2.2. Consider

$$H = \begin{bmatrix} s_1 - 1 & 0 & 0 \\ 0 & s_2 - 1 & 0 \\ 0 & 0 & s_3 - 1 \\ 1 & 1 & 1 \\ 1 & 2 & 3 \end{bmatrix} \tag{5.17}$$

whose determinantal ideal $\mathcal{I}(H) = \langle s_1 - 1, s_2 - 1, s_3 - 1 \rangle$ has co-dimension three. Hence H does not possess a minimal left annihilator of full row rank.

Now we turn to the announced example of a minimal ON representation that is not observable.

Example 5.2.3. Consider

$$\mathcal{B} = \{ w : \mathbb{Z}^2 \to \mathbb{R}, \ (\sigma_1 - 1)(\sigma_2 - 1)w = 0 \}.$$

An ON representation of dimension two is given by

$$\begin{bmatrix} \sigma_1 & 0 \\ 0 & \sigma_2 \end{bmatrix} \begin{bmatrix} x_1 \\ x_2 \end{bmatrix} = \begin{bmatrix} 1 & -1 \\ 0 & 1 \end{bmatrix} \begin{bmatrix} x_1 \\ x_2 \end{bmatrix} \tag{5.18}$$

$$0 = x_1 + x_2 - w. \tag{5.19}$$

This can be seen as follows: A minimal left annihilator of

$$H = \begin{bmatrix} s_1 - 1 & 1 \\ 0 & s_2 - 1 \\ 1 & 1 \end{bmatrix}$$

is given by $X = \begin{bmatrix} -(s_2 - 1), & -(s_1 - 2), & (s_1 - 1)(s_2 - 1) \end{bmatrix}$, hence

$$\exists x : \mathbb{Z}^2 \to \mathbb{R}^2 : H(\sigma_1, \sigma_2)x = \begin{bmatrix} 0 \\ 0 \\ 1 \end{bmatrix} w \quad \Leftrightarrow \quad (\sigma_1 - 1)(\sigma_2 - 1)w = 0.$$

The ON representation (5.18), (5.19) is certainly state-minimal and even of least order. On the other hand, H is not zero right prime as its rank drops at $\lambda = (2, 1)$, and thus the 2×2 minors of H have a common zero there, contradicting condition 3 in Lemma 5.2.2.

Next, we turn to the question of reducibility of ON representations to standard Roesser-type input-state-output models. First, we introduce a suitable notion of inputs and outputs.

Definition 32 *[48] For a behavior* $\mathcal{B} = \{w : \mathbb{Z}^r \to \mathbb{F}^q, R(\sigma, \sigma^{-1})w = 0\}$, *with* $R \in \mathbb{F}[s, s^{-1}]^{g \times q}$, *an* **input-output structure** *is a partition*

$$R\Pi = \begin{bmatrix} -Q & P \end{bmatrix} \quad \text{with } P \in \mathbb{F}[s, s^{-1}]^{g \times p} \text{ and } \operatorname{rank}(P) = \operatorname{rank}(R) = p,$$

where Π *is a permutation matrix. Then there exists a uniquely determined rational matrix* $G \in \mathbb{F}(s)^{p \times (q-p)}$ *such that* $Q = PG$. *It is called the* **transfer matrix** *of* $\mathcal{B}$ *with respect to the partition*

$$w = \Pi \begin{bmatrix} u \\ y \end{bmatrix}.$$

Setting $m := q - p$, *one can rewrite the behavior as*

$$\mathcal{B} = \Pi \left\{ \begin{bmatrix} u \\ y \end{bmatrix} : \mathbb{Z}^r \to \mathbb{F}^m \times \mathbb{F}^p, \ P(\sigma, \sigma^{-1})y = Q(\sigma, \sigma^{-1})u \right\}. \tag{5.20}$$

Note that every behavior possesses an input-output structure, since it is always possible to select $p = \operatorname{rank}(R)$ linearly independent (over $\mathbb{F}(s)$) columns of R. The notion "input" is justified by the fact that u is a vector of free variables in the sense that for all $u : \mathbb{Z}^r \to \mathbb{F}^m$, there exists $y : \mathbb{Z}^r \to \mathbb{F}^p$ such that [48]

$$w = \Pi \begin{bmatrix} u \\ y \end{bmatrix} \in \mathcal{B}.$$

Once u is chosen, none of the components of y is free (more precisely, the zero–input behavior is autonomous).

Let $R = X_1 B - X_2 D$ be a kernel representation derived from an ON representation according to (5.16). Let

$$R\Pi = \begin{bmatrix} X_1 & X_2 \end{bmatrix} \begin{bmatrix} B \\ -D \end{bmatrix} \Pi = \begin{bmatrix} X_1 & X_2 \end{bmatrix} \begin{bmatrix} B_u & B_y \\ -D_u & -D_y \end{bmatrix}$$

be an input–output structure of $\mathcal{B} = \ker(R)$. Then with the corresponding partition

$$\Pi^{-1}w =: \begin{bmatrix} u \\ y \end{bmatrix}$$

the ON representation takes the form

$$\begin{aligned} \Delta(\sigma)x &= Ax + B_u u + B_y y \\ 0 &= Cx + D_u u + D_y y. \end{aligned}$$

In particular, any behavior with an input-output structure (5.20) can be represented by ON equations of this type. If the second equation can be solved for y, this gives rise to an input-state-output (ISO) representation of $\mathcal{B}$, that is,

$$\begin{bmatrix} u \\ y \end{bmatrix} \in \mathcal{B} \quad \Leftrightarrow \quad \exists x \text{ with } \begin{cases} \Delta(\sigma)x &= \tilde{A}x + \tilde{B}u \\ y &= \tilde{C}x + \tilde{D}u. \end{cases}$$

Definition 33 *Let* $(A, B, C, D) \in \mathbb{F}(n, p, m)$ *be given and let* $n \in \mathbb{N}^r$ *be such that* $|n| = n$. *Define* $\Delta(s) = \mathrm{diag}(s_1 I_{n_1}, \dots, s_r I_{n_r})$. *The data* $(n; A, B, C, D)$ *or the equations*

$$\begin{aligned} \Delta(\sigma)x &= Ax + Bu & (5.21) \\ y &= Cx + Du & (5.22) \end{aligned}$$

are called an **input-state-output (ISO) representation** *of* $\mathcal{B}$ *if there exists a permutation matrix* Π *such that*

$$\mathcal{B} = \Pi \left\{ \begin{bmatrix} u \\ y \end{bmatrix} : \mathbb{Z}^r \to \mathbb{F}^{m+p}, \ \exists x : \mathbb{Z}^r \to \mathbb{F}^n \text{ such that } (5.21), (5.22) \text{ hold} \right\}.$$

In analogy to the two-dimensional situation, the representation given by (5.21), (5.22) is called a **Roesser model** [62]. The property of leading to such an ISO representation motivates the subsequent definition:

Definition 34 *An ON representation* $(n; A, B, C, D)$ *is said to be* **proper** *if*

$$\mathrm{im}(C) \subseteq \mathrm{im}(D).$$

Lemma 5.2.4. $\mathcal{B}$ *admits a proper ON representation iff* $\mathcal{B}$ *admits an ISO representation.*

Proof. An ISO representation $(n; A, B, C, D)$ can be rewritten as

$$\Delta(\sigma)x = Ax + \begin{bmatrix} B & 0 \end{bmatrix} \begin{bmatrix} u \\ y \end{bmatrix}$$

$$0 = Cx + \begin{bmatrix} D & -I \end{bmatrix} \begin{bmatrix} u \\ y \end{bmatrix},$$

which is a proper ON representation (without loss of generality, let $\Pi = I$).

Conversely, let $(n; A, B, C, D)$ be a proper ON representation. Without loss of generality, let $\begin{bmatrix} C & D \end{bmatrix}$ be a full row rank matrix. Hence properness of the ON representation amounts to D having full row rank. Then there exists a permutation matrix Π such that

$$D\Pi = \begin{bmatrix} D_u & D_y \end{bmatrix} \quad \text{with } D_y \in \mathbb{F}^{g \times g}, \quad \det(D_y) \neq 0.$$

Define

$$\begin{bmatrix} u \\ y \end{bmatrix} := \Pi^{-1}w \quad \text{and} \quad \begin{bmatrix} B_u & B_y \end{bmatrix} := B\Pi,$$

where the partitions correspond to that of $D\Pi$. Then

$$0 = Cx + Dw = Cx + D_u u + D_y y,$$

can be solved for y and thus

$$\Delta(\sigma)x = Ax + B_u u + B_y y = (A - B_y D_y^{-1}C)x + (B_u - B_y D_y^{-1}D_u)u$$

$$y = -D_y^{-1}Cx - D_y^{-1}D_u u$$

is an ISO representation of $\mathcal{B}$. □

As should be expected, an ISO representation of $\mathcal{B}$ gives rise to an input-output structure. Suppose (letting $\Pi = I$)

$$\begin{bmatrix} u \\ y \end{bmatrix} \in \mathcal{B} \quad \Leftrightarrow \quad \exists x \text{ with } H(\sigma)x := \begin{bmatrix} \Delta(\sigma) - A \\ C \end{bmatrix} x = \begin{bmatrix} B & 0 \\ -D & I \end{bmatrix} \begin{bmatrix} u \\ y \end{bmatrix}.$$

Let $X = \begin{bmatrix} X_1 & X_2 \end{bmatrix}$ be a minimal left annihilator (over $\mathbb{F}[s, s^{-1}]$) of H, then

$$\begin{bmatrix} u \\ y \end{bmatrix} \in \mathcal{B} \quad \Leftrightarrow \quad \begin{bmatrix} X_1 B - X_2 D, & X_2 \end{bmatrix} (\sigma, \sigma^{-1}) \begin{bmatrix} u \\ y \end{bmatrix} = 0. \qquad (5.23)$$

Set $P := X_2$, and $Q := -X_1 B + X_2 D$. From $X_1 + X_2 C(\Delta - A)^{-1} = 0$, we conclude that the columns of X_1 can be written as a linear combination (with rational coefficients) of the columns of X_2, hence

$$p = \operatorname{rank}(X) = \operatorname{rank}(X_2) = \operatorname{rank} \begin{bmatrix} X_1 B - X_2 D, & X_2 \end{bmatrix}.$$

As expected, the transfer matrix is

$$G = C(\Delta - A)^{-1} B + D.$$

The following definition should be compared with the corresponding one-dimensional notion [35, p. 384].

Definition 35 *Any Laurent polynomial matrix $R \in \mathbb{F}[s, s^{-1}]^{g \times q}$ can be uniquely written as*

$$R(s, s^{-1}) = S(s, s^{-1}) R_{hr} + L(s, s^{-1}),$$

where

$$S(s, s^{-1}) = \operatorname{diag}(s^{k_1}, \ldots, s^{k_g}) \in \mathbb{F}[s, s^{-1}]^{g \times g}, \qquad k_i = \vee(\operatorname{supp}(R_{i-})) \in \mathbb{Z}^r,$$

and $R_{hr} \in \mathbb{F}^{g \times q}$. The matrix L contains the "lower order" terms, that is, for every $t_i \in \operatorname{supp}(L_{i-}) \subset \mathbb{Z}^r$, we have

$$(t_i)_j \le (k_i)_j \text{ for } 1 \le j \le r \quad \text{and} \quad t_i \ne k_i.$$

*The matrix R is said to be **row-proper** (or **row-reduced**) if R_{hr}, the **highest–row–degree coefficient matrix** of R, is a full row rank matrix. Note that then also $\operatorname{rank}(R) = \operatorname{rank}(R_{hr}) = g$.*

*We say that R is **weakly row-proper** if*

1. $\operatorname{rank}(R) = \operatorname{rank}(R_{hr})$;

*2. R_{hr} contains no zero row, that is, each row of R is **cw-unital** [48], or*

$$\vee(\operatorname{supp}(R_{i-})) \in \operatorname{supp}(R_{i-}).$$

In this notion of row-properness, as well as in the type of ON representations considered here, the hyper-quadrant $\mathbb{N}^r$ of $\mathbb{Z}^r$ plays a special role. It should be kept in mind that one can proceed like this with any of the 2^r hyper-quadrants of $\mathbb{Z}^r$.

Comparing with (5.5), (5.6), it is easy to see that a row-reduced matrix $R = S R_{hr} + L$ and

$$\tilde{R} := S^{-1} R = R_{hr} + S^{-1} L$$

generate the same behavior, and $\tilde{R} \in \mathbb{F}[s^{-1}]^{g \times q}$ with $\vee(\operatorname{supp}(\tilde{R}_{i-})) = 0$ for all i. Moreover, $\tilde{R}$ admits a representation

$$\tilde{R} = R_{hr} + \sum_{0 \neq t \in \mathbb{N}^r} R(-t)s^{-t}, \qquad R(-t) \in \mathbb{F}^{g \times q}$$

with a full row rank "highest coefficient matrix" $R_{hr} =: \tilde{R}(0)$. Then we may write

$$\tilde{R}(s^{-1}) = C\Delta(s^{-1})(I - A\Delta(s^{-1}))^{-1}B + R_{hr}.$$

The following is an immediate consequence.

Corollary 16 *If a behavior admits a kernel representation $\mathcal{B} = \ker(R)$ with a row-proper matrix R, then it possesses a proper ON and hence an ISO representation.*

The converse is not true: A row-proper matrix has full row rank, but a behavior with an ISO representation does not necessarily admit a full row rank representation.

Example 5.2.4. Take $n = (1, 1, 1)$,

$$A = I_3, \quad B = I_3, \quad C = \begin{bmatrix} 1 & 1 & 1 \\ 1 & 2 & 3 \end{bmatrix}, \quad D = 0.$$

In view of (5.23), the kernel representations of the associated behavior are precisely the minimal left annihilators of H from (5.17). We have already noted that although $p = 2$, there is no MLA of H with only two rows.

However, we have the following weaker result.

Theorem 23 *If $\mathcal{B}$ possesses an ISO representation, then it admits an autoregressive representation with a weakly row-proper kernel representation matrix.*

Proof. It is easy to check that

$$\tilde{X} := [\ -C\Delta^{-1}\mathrm{adj}\,(I - A\Delta^{-1}), \quad \det(I - A\Delta^{-1})I_p \ ] \in \mathbb{F}[s^{-1}]^{p \times (n+p)}$$

is a left annihilator of

$$\tilde{H} := H\Delta^{-1} = \begin{bmatrix} I - A\Delta^{-1} \\ C\Delta^{-1} \end{bmatrix} \in \mathbb{F}[s^{-1}]^{(n+p) \times n}.$$

Let $X(s^{-1}) = [\ X_1(s^{-1}), \quad X_2(s^{-1}) \ ]$ be a MLA of $\tilde{H}$ over $\mathbb{F}[s^{-1}]$. Note that then it is also a MLA of $\tilde{H}$ (and thus of H) over $\mathbb{F}[s, s^{-1}]$. There exists a polynomial matrix Z, with entries in $\mathbb{F}[s^{-1}]$, such that $\tilde{X} = ZX$. In particular,

$$\det(I - A\Delta(s^{-1}))I_p = Z(s^{-1})X_2(s^{-1}).$$

Thus $I_p = Z(0)X_2(0)$. We conclude that $\mathrm{rank}\,(X_2(0)) = p$. In view of (5.23),

$$R = [\ X_1(s^{-1})B - X_2(s^{-1})D,\ \ X_2(s^{-1})\ ]$$

is a kernel representation of $\mathcal{B}$. Without loss of generality, $X_2(0)$ contains no zero row (this can be achieved by pre-multiplication of R by a suitable non-singular constant matrix) and thus $\vee(\mathrm{supp}(R_{i-})) = 0$ for all i. Hence $R_{hr} = R(0)$ and R is weakly row-proper.   □

If X in the proof above can be chosen to be a full-row-rank matrix (this is always true for systems of dimension less than three), we even obtain a row-proper kernel representation.

5.3 Driving–variable Representations

Let $\mathcal{B} = \{w : \mathbf{Z}^r \to \mathbf{F}^q, \exists v : \mathbf{Z}^r \to \mathbf{F}^m : w = M(\sigma, \sigma^{-1})v\}$ be a behavior in image representation. Recall that such a behavior is necessarily controllable. We proceed with the matrix $M \in \mathbf{F}[s, s^{-1}]^{q \times m}$ as with R before: first, we shift its support into the non-positive hyper-quadrant by post-multiplying it by $\mathrm{diag}(s^{-k_1}, \ldots, s^{-k_m})$, where $k_i = \vee(\mathrm{supp}(M_{-i}))$. Then we can write, without loss of generality,

$$M = M(s^{-1}) = C(\Delta(s) - A)^{-1}B + D,$$

where A, B, C, D are constant matrices, $\Delta(s) = \mathrm{diag}(s_1 I_{n_1}, \ldots, s_r I_{n_r})$ and $\Delta(s) - A$ is invertible over the Laurent polynomial ring. We define, for each v,

$$x = (\Delta(\sigma) - A)^{-1}Bv,$$

then $w = Cx + Dv$ and hence

$$w \in \mathcal{B} \quad \Leftrightarrow \quad \exists x, v \text{ with } \begin{cases} \Delta(\sigma)x &=& Ax + Bv \\ w &=& Cx + Dv. \end{cases}$$

Definition 36 *For $n \in \mathbf{N}^r$, $n = |n|$, define $\Delta(s) = \mathrm{diag}(s_1 I_{n_1}, \ldots, s_r I_{n_r})$. Let $(A, B, C, D) \in \mathbf{F}(n, q, m)$ be given. The equations*

$$\Delta(\sigma)x \quad = \quad Ax + Bv \qquad (5.24)$$
$$w \quad = \quad Cx + Dv \qquad (5.25)$$

or equivalently, the data $(n; A, B, C, D)$, are called a **driving-variable (DV) representation** *of $\mathcal{B}$ if $\mathcal{B} = \pi_w \mathcal{B}_{DV}(n; A, B, C, D)$, where*

$$\mathcal{B}_{DV}(n; A, B, C, D) =$$

$$\{(x, w) : \mathbb{Z}^r \to \mathbb{F}^n \times \mathbb{F}^q, \exists v : \mathbb{Z}^r \to \mathbb{F}^m \text{ such that } (5.24), (5.25) \text{ hold}\}.$$

A DV representation $(n; A, B, C, D)$ of $\mathcal{B}$ as above is **minimal** if

$$\left. \begin{array}{l} \mathcal{B} = \pi_w \mathcal{B}_{DV}(n', A', B', C', D') \text{ for some } n' \in \mathbb{N}^r, \\ (A', B', C', D') \in \mathbb{F}(n', q, m'), \quad n' = |n'| \end{array} \right\} \quad (5.26)$$

implies that $n < n'$ or $m < m'$ or $(n, m) = (n', m')$. If the implication of (5.26) is even $n \leq n'$ and $m \leq m'$, then the DV representation is said to have **least order**. We say that a DV representation is **state-minimal** if the implication of (5.26) is only $n \leq n'$.

If $(n; A, B, C, D)$ is an DV representation of $\mathcal{B}$, then so is

$$(n; T^{-1}(A + BF)T, T^{-1}BQ, (C + DF)T, DQ),$$

there $T = \text{diag}(T_1, \dots, T_r)$, $T_i \in \mathbb{F}^{n_i \times n_i}$ and $Q \in \mathbb{F}^{m \times m}$ are non-singular, and F is an arbitrary $m \times n$ matrix over $\mathbb{F}$.

Definition 37 A DV representation $(n; A, B, C, D)$ is called **proper** if

$$\ker(D) \subseteq \ker(B).$$

Theorem 24 The following are equivalent: $\mathcal{B}$ possesses

1. a proper DV representation;
2. a proper ON representation;
3. an ISO representation.

Proof. In view of Lemma 5.2.4, it remains to show the equivalence of Assertions 1 and 3. An ISO representation $(n; A, B, C, D)$ can be written as

$$\Delta(\sigma)x = Ax + Bv$$

$$\begin{bmatrix} u \\ y \end{bmatrix} = \begin{bmatrix} 0 \\ C \end{bmatrix} x + \begin{bmatrix} I \\ D \end{bmatrix} v$$

which is a proper DV representation.

Conversely, let $(n; A, B, C, D)$ be a proper DV representation of $\mathcal{B}$. Without loss of generality, $\begin{bmatrix} B^T & D^T \end{bmatrix}^T$ has full column rank. Then D itself has full column rank, *i.e.*, there exists a permutation matrix Π such that

$$\Pi D = \begin{bmatrix} D_u \\ D_y \end{bmatrix} \quad \text{with } D_u \in \mathbb{F}^{m \times m}, \quad \det(D_u) \neq 0.$$

Define

$$\begin{bmatrix} u \\ y \end{bmatrix} := \Pi w \quad \text{and} \quad \begin{bmatrix} C_u \\ C_y \end{bmatrix} := \Pi C,$$

where the partitions correspond to that of ΠD. Pre-multiply $w = Cx + Dv$
by Π to obtain

$$\begin{aligned} u &= C_u x + D_u v \\ y &= C_y x + D_y v. \end{aligned}$$

Hence $v = D_u^{-1}(u - C_u x)$. Plugging in yields the desired ISO representation
of $\mathcal{B}$:

$$\begin{aligned} \Delta(\sigma)x &= (A - BD_u^{-1}C_u)x + BD_u^{-1}u \\ y &= (C_y - D_y D_u^{-1}C_u)x + D_y D_u^{-1}u. \end{aligned}$$

$\square$

It is sometimes convenient to pass from a proper ON to a proper DV
representation without making the detour to ISO representations. This is
accomplished as follows:

Let $(n; A, B, C, D)$ be a proper ON representation. Then there exists $C' \in \mathbb{F}^{q \times n}$ such that $C = DC'$. Then the output-nulling equations take the form
$0 = D(C'x + w)$. Setting $v' := C'x + w$, the ON representation is equivalent
to

$$\Delta(\sigma)x = (A - BC')x + Bv', \quad w = -C'x + v', \quad v' \in \ker(D).$$

Let $m := q - \mathrm{rank}\,(D)$ and let $\ker(D) \subseteq \mathbb{F}^q$ be generated by the columns
of $D' \in \mathbb{F}^{q \times m}$, i.e., $\ker(D) = \mathrm{im}(D')$ for $D' : \mathbb{F}^m \to \mathbb{F}^q$. In other words,
D' is a minimal right annihilator of D. We obtain the following proper DV
representation of $\mathcal{B}$:

$$\begin{aligned} \Delta(\sigma)x &= (A - BC')x + BD'v \\ w &= -C'x + D'v. \end{aligned}$$

Conversely, in a proper DV representation, there exists B' such that $B = B'D$. Let D' be a minimal left annihilator (MLA) of D, then

$$\begin{bmatrix} I & -B' \\ 0 & D' \end{bmatrix} \quad \text{is a MLA of} \quad \begin{bmatrix} B \\ D \end{bmatrix}.$$

Thus a necessary and sufficient condition for the existence of $v : \mathbb{Z}^r \to \mathbb{F}^m$
with

$$\begin{bmatrix} \Delta(\sigma) - A & 0 \\ -C & I \end{bmatrix} \begin{bmatrix} x \\ w \end{bmatrix} = \begin{bmatrix} B \\ D \end{bmatrix} v$$

is

$$\begin{bmatrix} \Delta(\sigma) - A + B'C & -B' \\ -D'C & D' \end{bmatrix} \begin{bmatrix} x \\ w \end{bmatrix} = \begin{bmatrix} 0 \\ 0 \end{bmatrix}$$

which constitutes a proper ON representation.

Definition 38 *A DV representation is called* **controllable** *if*

$$\mathcal{B}_{xv} = \{(x,v) : \mathbb{Z}^r \to \mathbb{F}^n \times \mathbb{F}^m, \; \Delta(\sigma)x = Ax + Bv\}$$

is controllable, i.e., if $\mathcal{B}_{xv}$ *possesses an image representation.*

Corollary 17

1. *Let* $(n; A, B, C, D)$ *be a DV representation. Define the Hautus matrix*

$$H := \begin{bmatrix} \Delta(s) - A, & B \end{bmatrix} \in \mathbb{F}[s, s^{-1}]^{n \times (n+m)}.$$

   The DV representation is controllable iff H *is GFLP over* $\mathbb{F}[s, s^{-1}]$.
2. $\mathcal{B}$ *is controllable iff it possesses a controllable DV representation.*

Proof. Assertion 1 follows from Chapter 1. If $\mathcal{B}$ possesses an image representation, then the construction of a DV representation described above yields a matrix H that is zero left prime, hence GFLP. Conversely, let $(n; A, B, C, D)$ be a controllable DV representation. Then for some integer $k \geq 1$, and some Laurent polynomial matrices N_1, N_2 of appropriate size,

$$\mathcal{B}_{xv} = \left\{ \begin{bmatrix} x \\ v \end{bmatrix} : \mathbb{Z}^r \to \mathbb{F}^{n+m}, \; \exists l : \mathbb{Z}^r \to \mathbb{F}^k : \begin{bmatrix} x \\ v \end{bmatrix} = \begin{bmatrix} N_1 \\ N_2 \end{bmatrix} (\sigma, \sigma^{-1})l \right\}$$

and

$$\mathcal{B} = \{w : \mathbb{Z}^r \to \mathbb{F}^q, \; \exists l : \mathbb{Z}^r \to \mathbb{F}^k, \; w = (CN_1 + DN_2)(\sigma, \sigma^{-1})l\},$$

i.e., $\mathcal{B}$ possesses an image representation. □

In this chapter, a new class of first order representations of linear constant–coefficient systems of partial difference equations has been proposed: It resembles the Roesser model, but it does not require causality, which is an unnecessary restriction in the multivariate situation. Still, it features regular (*i.e.*, explicit) updating equations, which are convenient for practical implementation. Minimality issues have been addressed, but are still far from being fully understood. The question of reducibility to a standard input-state-output setting has been discussed; necessary and sufficient conditions have been given. For the class of controllable systems, an alternative first order scheme is available, which is the dual counterpart of the above-mentioned representations.

6. Linear Fractional Transformations

A mapping of the form

$$F : \mathbb{C} \setminus \{s,\ \alpha s + \beta = 0\} \longrightarrow \mathbb{C}, \quad s \longmapsto \frac{\gamma s + \delta}{\alpha s + \beta},$$

where $\alpha, \beta, \gamma, \delta \in \mathbb{C}$ and $(\alpha, \beta) \neq (0,0)$, is called a **linear fractional transformation (LFT)**. If $\beta \neq 0$, it can be rewritten as

$$F : \mathbb{C} \setminus \{s,\ 1 - as = 0\} \longrightarrow \mathbb{C}, \quad s \longmapsto d + cs(1 - as)^{-1}, \qquad (6.1)$$

where

$$a = -\frac{\alpha}{\beta}, \quad c = \frac{\beta\gamma - \alpha\delta}{\beta^2}, \quad d = \frac{\delta}{\beta}.$$

Equation (6.1) motivates the following generalization to the matrix case:

$$F : \mathbb{C}^{n \times n} \setminus \{\Delta,\ \det(I - A\Delta) = 0\} \to \mathbb{C}^{g \times q}, \quad \Delta \mapsto D + C\Delta(I - A\Delta)^{-1}B,$$

where A, B, C, D are complex matrices of dimensions $n \times n$, $n \times q$, $g \times n$, $g \times q$, respectively. Taking explicit account of the coefficient matrices, one writes

$$F(\Delta) = \mathcal{F}(\Delta; A, B, C, D) = D + C\Delta(I - A\Delta)^{-1}B.$$

The following interpretation in terms of a feedback loop is often useful.

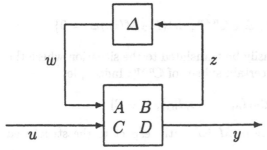

Fig. 6.1. Feedback Interpretation of Linear Fractional Transformation.

The constituent equations of the loop are

$$\begin{bmatrix} z \\ y \end{bmatrix} = \begin{bmatrix} A & B \\ C & D \end{bmatrix} \begin{bmatrix} w \\ u \end{bmatrix} \quad \text{and} \quad w = \Delta z.$$

Hence $(I - A\Delta)z = Bu$. Assuming well-posedness of the loop, *i.e.*, invertibility of $I - A\Delta$, we can solve for z and hence

$$y = (D + C\Delta(I - A\Delta)^{-1}B)u,$$

that is, the LFT can be seen as the "closed loop transfer function from u to y."

Linear fractional transformations are prominent in H_∞ and robust control [12, 97]. Let M be a proper real-rational and stable (1D) transfer matrix, *i.e.*, the entries of M are of the form $\frac{n}{d}$ with polynomials n, d and $\deg(n) \leq \deg(d)$, moreover, M has no poles in the closed right half of the complex plane. Then the H_∞ norm of M is defined as

$$\|M\|_\infty := \sup_{\omega \in \mathbf{R}} \|M(i\omega)\|_2.$$

The **Small Gain Theorem** [97, p. 217] says that, for a given proper and stable

$$M = \begin{bmatrix} M_{11} & M_{12} \\ M_{21} & M_{22} \end{bmatrix},$$

the condition $\|M_{11}\|_\infty < \gamma$ is necessary and sufficient for the linear fractional transformation

$$F(\Delta) = M_{22} + M_{21}\Delta(I - M_{11}\Delta)^{-1}M_{12} \tag{6.2}$$

to exist and to be proper and stable for all complex matrices Δ with $\|\Delta\|_2 \leq \gamma^{-1}$. This robust stability result has been refined by Packard and Doyle [11, 50] in order to take account of additional information on the structure of Δ. Note that

$$\|M(i\omega)\|_2 = \max\{\|\Delta\|_2^{-1}, \Delta \in \mathbb{C}^{n \times n}, \det(I - M(i\omega)\Delta) = 0\}.$$

Now this characterization can easily be translated to the situation where the uncertainty Δ is restricted to a certain subset of $\mathbb{C}^{n \times n}$. Indeed, let

$$\mathbf{\Delta} = \{\Delta = \operatorname{diag}(\delta_1 I_{n_1}, \dots, \delta_r I_{n_r}), \delta_i \in \mathbb{C}\}.$$

The **structured singular value** of $M(i\omega)$ with respect to the structured set $\mathbf{\Delta}$ is

$$\mu_{\mathbf{\Delta}}(M(i\omega)) = \max\{\|\Delta\|_2^{-1}, \Delta \in \mathbf{\Delta}, \det(I - M(i\omega)\Delta) = 0\}$$

unless there is no such $\Delta \in \pmb{\Delta}$, in which case we set $\mu_{\pmb{\Delta}}(M(i\omega)) = 0$. Then (6.2) exists and is proper and stable for all $\Delta \in \pmb{\Delta}$ with $\|\Delta\|_2 \leq \gamma^{-1}$ iff $\sup_{\omega \in \mathbb{R}} \mu_{\pmb{\Delta}}(M(i\omega)) < \gamma$.

For robust stability issues, it is therefore desirable to represent an uncertain transfer matrix in terms of an LFT.

Extraction of parametric uncertainty: Consider a transfer matrix P that depends rationally on s and on several parametric uncertainties δ_i,

$$P \in \mathbb{R}(s, \delta_1, \dots, \delta_r)^{p \times m}.$$

Suppose that P is **causal** [30] with respect to $\delta = (\delta_1, \dots, \delta_r)$, that is, each entry takes the form

$$\frac{\sum_{n \in \mathbb{N}^r} a(s, n)\delta^n}{\sum_{n \in \mathbb{N}^r} b(s, n)\delta^n}$$

with $b(s, 0, \dots, 0) \neq 0$. As the δ_i usually model a variation of some parameter

$$p_i = v_i + m_i \delta_i$$

around a nominal value v_i, there is no loss of generality in this causality requirement, because a reasonably chosen nominal point will guarantee that $P(s, 0, \dots, 0)$ is well-defined. Then P has a representation in terms of an LFT

$$P = M_{22} + M_{21}\Delta(I - M_{11}\Delta)^{-1}M_{12},$$

where $\Delta = \mathrm{diag}(\delta_1 I_{n_1}, \dots, \delta_r I_{n_r})$ for some suitably chosen integers $n_1, \dots, n_r$. The matrices M_{ij} are rational functions of s alone. The dependencies of P on s and δ are thus separated from each other; the uncertainty (Δ) enters a nominal system (M) in terms of feedback loop as above. One says that the uncertainty has been extracted.

An important feature of linear fractional transformations is that "the LFT of an LFT is again an LFT." This implies that for parametrized realizations

$$
\begin{aligned}
sx &= A(\delta)x + B(\delta)u \\
y &= C(\delta)x + D(\delta)u,
\end{aligned}
$$

it suffices to write the coefficient matrix

$$\begin{bmatrix} A(\delta) & B(\delta) \\ C(\delta) & D(\delta) \end{bmatrix}$$

as an LFT, from which an LFT representation of $D + C(sI - A)^{-1}B = F(s^{-1}I)$ is easily available. A similar argument holds for descriptor systems [72, 93]

$$E(\delta)sx \;=\; A(\delta)x + B(\delta)u$$
$$y \;=\; C(\delta)x + D(\delta)u$$

and general polynomial systems (Rosenbrock systems) [92]

$$T(s,\delta)x \;=\; U(s,\delta)u$$
$$y \;=\; V(s,\delta)x + W(s,\delta)u.$$

For details, see Section 6.4.

6.1 LFT Construction

We consider the problem of constructing an LFT representation of a polynomial matrix $M(\delta) \in R[\delta]^{g \times q}$, where $\delta = (\delta_1,\dots,\delta_r)$ and R denotes an arbitrary commutative coefficient ring with unity. The main cases that should be thought of are $R = \mathbb{R}$, $\mathbb{R}[s]$, $\mathbb{R}[s,s^{-1}]$ and $\mathbb{R}(s)$.

Given $M(\delta) \in R[\delta]^{g \times q}$, we wish to construct integers $n_1,\dots,n_r \geq 0$, $n = \sum n_i$, and matrices M, Δ with

$$\Delta = \mathrm{diag}(\delta_1 I_{n_1},\dots,\delta_r I_{n_r}),$$

$$M = \begin{bmatrix} M_{11} & M_{12} \\ M_{21} & M_{22} \end{bmatrix} \in R^{(n+g)\times(n+q)}$$

such that

$$M(\delta) = \mathcal{F}(\Delta; M) := M_{22} + M_{21}\Delta(I - M_{11}\Delta)^{-1}M_{12}.$$

First, consider the case $q = 1$. Then $M(\delta)$ is merely a column vector of polynomials,

$$M(\delta) = \begin{bmatrix} M(\delta)_1 \\ \vdots \\ M(\delta)_g \end{bmatrix} = \sum_{i=1}^{g}\sum_{n\in\mathbb{N}^r} c_i(n)\delta^n e_i$$

where e_i denotes the i-th natural basis vector of R^g and $c_i(n) \in R$. Define the support of this polynomial column vector by

$$N := \mathrm{supp}(M(\delta)) = \{n \in \mathbb{N}^r, \exists i \text{ with } c_i(n) \neq 0\},$$

which is a finite subset of $\mathbb{N}^r$, and let $\delta^N = \{\delta^n,\, n \in N\}$. It is assumed that $M(\delta) \not\equiv 0$, i.e., $N \neq \emptyset$. As usual, let $|n| = \sum_{i=1}^{r} n_i$ for $n \in \mathbb{N}^r$. Define a directed graph Γ with vertices in the set of monic monomials $\{\delta^n,\, n \in \mathbb{N}^r\}$ in the following fashion: For each $n \in N$, choose a directed path from $1 = \delta^0$ to δ^n,

$$1 = v_0^{(n)} \to v_1^{(n)} \to v_2^{(n)} \to \ldots \to v_{|n|}^{(n)} = \delta^n \tag{6.3}$$

such that for all $0 \le i \le |n|$, there exists $j_i \in [r] := \{1, \ldots, r\}$ with

$$v_i^{(n)} = \delta_{j_i} v_{i-1}^{(n)}. \tag{6.4}$$

The vertex set of Γ is

$$V = \bigcup_{n \in N} \{v_0^{(n)}, \ldots, v_{|n|}^{(n)}\},$$

and each pair $(z, v) := (v_{i-1}^{(n)}, v_i^{(n)}) \in V^2$ according to (6.3), (6.4) above constitutes an edge of Γ with initial vertex z and terminal vertex v. The edge set is denoted by E.

By construction, $u := \delta^0 = 1 \in V$ is a **root** of the graph Γ, i.e., there exist directed paths from u to all the remaining vertices of Γ. In particular, Γ is quasi-strongly connected, which implies that it possesses a directed spanning tree [74, p. 106]. Without loss of generality, assume that Γ itself is a directed tree (otherwise, one can remove successively all the edges from Γ whose removal does not destroy the property of u being a root of Γ).

A systematic method for the direct construction of such a tree is the following: For each $v = \delta^n$ with $n \in N$, if $m \in N^r$ is such that

$$\exists j \in [r] \quad \text{with} \quad m_j \le n_j \quad \text{and} \quad m_i = \begin{cases} n_i & \text{for } i < j \\ 0 & \text{for } i > j, \end{cases}$$

then $\delta^m \in V$. For a given $n = (n_1, \ldots, n_r) \in N$, this means that the following predecessors of n (with respect to the lexicographic order) are exponents of elements in V:

$$(0, 0, \ldots, 0), (1, 0, \ldots, 0), \ldots, (n_1, 0, \ldots, 0),$$
$$(n_1, 1, 0, \ldots, 0), \ldots, (n_1, n_2, 0, \ldots, 0), \ldots,$$
$$(n_1, n_2, \ldots, n_{r-1}, 1), \ldots, (n_1, n_2, \ldots, n_r).$$

In a directed tree with root u, there exists, for every $v \in V \setminus \{u\}$, one and only one vertex $z(v) \in V$ such that $(z(v), v) \in E$. In other words, with $W := V \setminus \{u\}$, the map

$$W \to V, \quad w \mapsto z(w)$$

which assigns to each $w \in W$ its unique direct predecessor $z(w)$ is well-defined. It gives rise to a decomposition of W into a disjoint union

$$W = \bigcup_{i=1}^{r} {}^{\cdot} W_i \quad \text{with} \quad W_i = \{w \in W, w = \delta_i z(w)\}.$$

The elements of W are labeled with integers from the set $\{1, \ldots, n\}$, $n := |W|$, such that

$$w_i \in W_j \text{ and } w_k \in W_l \text{ with } j < l \quad \Rightarrow \quad i < k.$$

Then by defining vectors $w = (w_1, \dots, w_n)^T$ and, correspondingly, $z = (z_1, \dots, z_n)^T$ with $z_i = z(w_i)$, one obtains $w = \Delta z$ with

$$\Delta = \text{diag}(\delta_1 I_{n_1}, \dots, \delta_r I_{n_r}),$$

where $n_i = |W_i|$. Moreover, since each $z_i \in V = W \cup \{u\}$, there exist matrices $M_{11} \in \{0,1\}^{n \times n}$ and $M_{12} \in \{0,1\}^{n \times 1}$ such that

$$z = M_{11} w + M_{12} u.$$

Since $\delta^N \subseteq V$, where $N = \text{supp}(M(\delta))$, there exist matrices $M_{21} \in R^{g \times n}$ and $M_{22} \in R^{g \times 1}$ with

$$M(\delta) = M_{21} w + M_{22} u.$$

Altogether,

$$w = \Delta z \quad \text{and} \quad \begin{bmatrix} z \\ M(\delta) \end{bmatrix} = \begin{bmatrix} M_{11} & M_{12} \\ M_{21} & M_{22} \end{bmatrix} \begin{bmatrix} w \\ u \end{bmatrix}.$$

Thus, $z = M_{11} w + M_{12} u = M_{11} \Delta z + M_{12} u$. Using the fact that $I - M_{11} \Delta$ is invertible (see below), this yields $z = (I - M_{11} \Delta)^{-1} M_{12} u$ and

$$
\begin{aligned}
M(\delta) &= M_{22} u + M_{21} w \\
&= M_{22} u + M_{21} \Delta z \\
&= M_{22} u + M_{21} \Delta (I - M_{11} \Delta)^{-1} M_{12} u.
\end{aligned}
$$

Taking into account that $u = 1$, this yields the desired representation of $M(\delta)$ in terms of an LFT, namely

$$M(\delta) = \mathcal{F}(\Delta; M), \quad M = \begin{bmatrix} M_{11} & M_{12} \\ M_{21} & M_{22} \end{bmatrix} \in R^{(n+g) \times (n+q)}.$$

The case $q > 1$ can be reduced to the preceding construction method for $q = 1$ by an idealization technique: Given $M(\delta) \in R[\delta]^{g \times q}$, introduce a vector of q new indeterminates $u = (u_1, \dots, u_q)^T$ and consider the polynomial vector

$$M(\delta) u \in (R[\delta] u_1 \oplus \cdots \oplus R[\delta] u_q)^g \subset R[\delta, u]^g.$$

Write

$$M(\delta) u = \sum_{n \in N^r} \sum_{j=1}^{q} \sum_{i=1}^{g} c_i(n; j) \delta^n u_j e_i$$

with coefficients $c_i(n; j) \in R$, and define

$$N := \text{supp}(M(\delta)u) = \{(n; j) \in \mathbb{N}^r \times [q], \exists i \text{ with } c_i(n; j) \neq 0\}.$$

Define a directed graph Γ with vertex set $V \subset \{\delta^n u_j, (n; j) \in \mathbb{N}^r \times [q]\}$ by choosing paths

$$u_j = v_0^{(n)} \to v_1^{(n)} \to v_2^{(n)} \to \dots \to v_{|n|}^{(n)} = \delta^n u_j \tag{6.5}$$

from u_j to $\delta^n u_j$ for each $(n; j) \in N$ such that for all i,

$$v_i^{(n)} = \delta_{j_i} v_{i-1}^{(n)}. \tag{6.6}$$

The vertex set V of Γ consists of all $v_i^{(n)}$ according to (6.5), (6.6), and each pair $(v_{i-1}^{(n)}, v_i^{(n)}) \in V^2$ constitutes an edge of the graph. After possible removal of redundant edges, Γ becomes a directed forest with roots $u_1, \dots, u_q$ (assuming that $M(\delta)$ contains no zero column). Each component of the graph corresponds to a column of the matrix. Let $U \subseteq V$ be the set of roots of Γ. The predecessor map

$$W := V \setminus U \to V, \quad w \mapsto z(w)$$

gives rise to a representation $w = \Delta z$ with

$$\Delta = (\delta_1 I_{n_1}, \dots, \delta_r I_{n_r}),$$

where

$$n_i = |\{w \in W, w = \delta_i z(w)\}|,$$

$n = |W|$. The vectors $w := (w_1, \dots, w_n)^T$ and $z = (z_1, \dots, z_n)^T$ with $z_i := z(w_i)$ are supposed to be suitably sorted. Then

$$z = M_{11} w + M_{12} u$$

for some matrices $M_{11} \in \{0,1\}^{n \times n}$, $M_{12} \in \{0,1\}^{n \times q}$, and

$$M(\delta)u = M_{21} w + M_{22} u$$

for some matrices $M_{21} \in R^{g \times n}$ and $M_{22} \in R^{g \times q}$. Altogether, since $I - M_{11}\Delta$ is invertible according to the subsequent lemma,

$$M(\delta)u = (M_{22} + M_{21}\Delta(I - M_{11}\Delta)^{-1}M_{12})u = \mathcal{F}(\Delta; M)u,$$

where

$$M = \begin{bmatrix} M_{11} & M_{12} \\ M_{21} & M_{22} \end{bmatrix} \in R^{(n+g)\times(n+q)}.$$

This implies that $M(\delta) = \mathcal{F}(\Delta; M)$. The LFT construction is illustrated by the following example.

Example 6.1.1. (compare [72]): Consider

$$M(\delta) = \begin{bmatrix} \delta_1^2 & \delta_1 \delta_2 \\ \delta_2 \delta_3^2 & \delta_2 + \delta_3^2 \end{bmatrix} \in \mathbb{R}[\delta_1, \delta_2, \delta_3]^{2 \times 2}.$$

Introducing $u = (u_1, u_2)^T$, the support of $M(\delta)u$ is

$$N = \{(2,0,0;1), (0,1,2;1), (1,1,0;2), (0,1,0;2), (0,0,2;2)\}$$

corresponding to the vertices $\delta_1^2 u_1, \delta_2 \delta_3^2 u_1, \delta_1 \delta_2 u_2, \delta_2 u_2, \delta_3^2 u_2$ of the graph to be constructed.

A graph which satisfies the requirements above is represented in the following figure. Lattice points which correspond to elements of V are marked by circles, the big disks indicate the elements of the support of $M(\delta)u$.

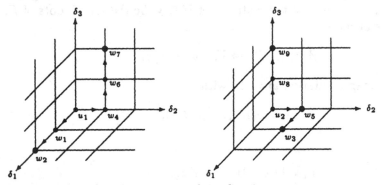

Fig. 6.2. The Two Components of the Graph.

The vertex set V consists of $w_1, \ldots, w_9$ and the two roots u_1, u_2. The vertices w_i and their corresponding direct predecessors z_i are given by

$$
\begin{aligned}
w_1 &:= & & \delta_1 u_1 & z_1 &= u_1 \\
w_2 &:= \delta_1^2 u_1 &= \delta_1 w_1 & & z_2 &= w_1 \\
w_3 &:= \delta_1 \delta_2 u_2 &= \delta_1 w_5 & & z_3 &= w_5 \\
w_4 &:= & & \delta_2 u_1 & z_4 &= u_1 \\
w_5 &:= & & \delta_2 u_2 & z_5 &= u_2 \\
w_6 &:= \delta_2 \delta_3 u_1 &= \delta_3 w_4 & & z_6 &= w_4 \\
w_7 &:= \delta_2 \delta_3^2 u_1 &= \delta_3 w_6 & & z_7 &= w_6 \\
w_8 &:= & & \delta_3 u_2 & z_8 &= u_2 \\
w_9 &:= \delta_3^2 u_2 &= \delta_3 w_8 & & z_9 &= w_8.
\end{aligned}
$$

The vertices have been labeled such that

$$W_1 = \{w_1, w_2, w_3\}, \quad W_2 = \{w_4, w_5\}, \quad W_3 = \{w_6, w_7, w_8, w_9\},$$

which gives rise to a partition of Δ as

$$\Delta = \mathrm{diag}(\delta_1 I_3,\ \delta_2 I_2,\ \delta_3 I_4).$$

The following equations yield the coefficient matrix M:

$$
\begin{bmatrix}
z_1 \\
z_2 \\
z_3 \\
z_4 \\
z_5 \\
z_6 \\
z_7 \\
z_8 \\
z_9 \\
\delta_1^2 u_1 + \delta_1 \delta_2 u_2 \\
\delta_2 \delta_3^2 u_1 + (\delta_2 + \delta_3^2) u_2
\end{bmatrix}
=
\left[
\begin{array}{ccc|cc|cccc||cc}
0 & 0 & 0 & 0 & 0 & 0 & 0 & 0 & 0 & 1 & 0 \\
1 & 0 & 0 & 0 & 0 & 0 & 0 & 0 & 0 & 0 & 0 \\
0 & 0 & 0 & 0 & 1 & 0 & 0 & 0 & 0 & 0 & 0 \\
0 & 0 & 0 & 0 & 0 & 0 & 0 & 0 & 0 & 1 & 0 \\
0 & 0 & 0 & 0 & 0 & 0 & 0 & 0 & 0 & 0 & 1 \\
0 & 0 & 0 & 1 & 0 & 0 & 0 & 0 & 0 & 0 & 0 \\
0 & 0 & 0 & 0 & 0 & 1 & 0 & 0 & 0 & 0 & 0 \\
0 & 0 & 0 & 0 & 0 & 0 & 0 & 0 & 0 & 0 & 1 \\
0 & 0 & 0 & 0 & 0 & 0 & 0 & 1 & 0 & 0 & 0 \\
\hline
\hline
0 & 1 & 1 & 0 & 0 & 0 & 0 & 0 & 0 & 0 & 0 \\
0 & 0 & 0 & 0 & 1 & 0 & 1 & 0 & 1 & 0 & 0
\end{array}
\right]
\begin{bmatrix}
w_1 \\
w_2 \\
w_3 \\
w_4 \\
w_5 \\
w_6 \\
w_7 \\
w_8 \\
w_9 \\
u_1 \\
u_2
\end{bmatrix}
$$

Single lines denote the partition that corresponds to the block structure of Δ, double lines separate the sub-matrices M_{ij}.

The graph-theoretic LFT construction above also yields a simple proof for the well-posedness of the resulting linear fractional transformations; compare [19].

Theorem 25 *In an LFT constructed as above, we have* $\det(I - M_{11}\Delta) = 1$.

Proof. In an acyclic directed graph with ρ vertices, there exists a **topological sorting** [74, p. 117], that is, the vertices can be labeled with integers from the set $\{1, \ldots, \rho\}$ such that

$$v_i \text{ predecessor of } v_j \quad \Rightarrow \quad i < j.$$

The vertex set of Γ is $V = U \dot\cup W$ with $U \subseteq \{u_1, \ldots, u_q\}$ ($U = \{u_1, \ldots u_q\}$ whenever $M(\delta)$ contains no zero column). Let $n = |W|$ and let Π be a permutation matrix which transforms the given vertex vector $v = \begin{bmatrix} w \\ u \end{bmatrix}$, $w_i \in W$, $u_j \in U$, into a topologically sorted one by

$$\tilde{v} = \Pi v$$

with $\tilde{v}$ being topologically sorted. Note that since $u_1, \ldots, u_q$ are vertices with zero in-degree (*i.e.*, no edge ends in one of the u_i), Π can, without loss of generality, be assumed to be of the form

$$\Pi = \begin{bmatrix} 0 & I \\ \Pi_1 & 0 \end{bmatrix}$$

with an $n \times n$ permutation matrix Π_1, *i.e.*,

$$\tilde{v} = \begin{bmatrix} 0 & I \\ \Pi_1 & 0 \end{bmatrix} \begin{bmatrix} w \\ u \end{bmatrix} = \begin{bmatrix} u \\ \Pi_1 w \end{bmatrix} =: \begin{bmatrix} u \\ \tilde{w} \end{bmatrix}.$$

Then with $\tilde{z} = \Pi_1 z$, $\tilde{w} = \Pi_1 w = \Pi_1 \Delta z = \Pi_1 \Delta \Pi_1^{-1} \tilde{z}$ and

$$
\begin{aligned}
\tilde{z} &= \Pi_1 \begin{bmatrix} M_{11} & M_{12} \end{bmatrix} v \\
&= \Pi_1 \begin{bmatrix} M_{11} & M_{12} \end{bmatrix} \Pi^{-1} \tilde{v} \\
&= \Pi_1 \begin{bmatrix} M_{11} & M_{12} \end{bmatrix} \begin{bmatrix} 0 & \Pi_1^{-1} \\ I & 0 \end{bmatrix} \begin{bmatrix} u \\ \tilde{w} \end{bmatrix} \\
&= \Pi_1 M_{12} u + \Pi_1 M_{11} \Pi_1^{-1} \tilde{w}.
\end{aligned}
$$

If $\tilde{z}_i = \tilde{w}_j$, i.e, if vertex $\tilde{w}_j$ is the predecessor of vertex $\tilde{w}_i$, then $j < i$ because of the topological sorting. Hence

$$\tilde{M}_{11} := \Pi_1 M_{11} \Pi_1^{-1} = \begin{bmatrix} 0 & & 0 \\ & \ddots & \\ * & & 0 \end{bmatrix},$$

that is, $(\tilde{M}_{11})_{ij} = 0$ for $j \geq i$. This implies that

$$\Pi_1 (I - M_{11}\Delta) \Pi_1^{-1} = I - \tilde{M}_{11}(\Pi_1 \Delta \Pi_1^{-1}) = \begin{bmatrix} 1 & & 0 \\ & \ddots & \\ * & & 1 \end{bmatrix}$$

and hence $\det(I - M_{11}\Delta) = 1$.   □

6.2 Reduction to Trim Form

Definition 39 *Let $n = (n_1, \dots, n_r)$, $n = |n|$, and*

$$\begin{bmatrix} A & B \\ C & D \end{bmatrix} \in \mathbb{F}^{(n+g) \times (n+q)}$$

be given, where $\mathbb{F}$ is an arbitrary field, and consider the LFT defined by $(n; A, B, C, D)$, that is,

$$\mathcal{F}(n; A, B, C, D) := D + C\Delta(I - A\Delta)^{-1}B,$$

*where $\Delta = \mathrm{diag}(\delta_1 I_{n_1}, \dots, \delta_r I_{n_r})$. The number n is called **state-dimension** of the LFT representation and $(n; A, B, C, D)$ is said to be **minimal** if its state-dimension is minimal among all the LFT representations of a matrix, i.e., if*

$$\mathcal{F}(n; A, B, C, D) = \mathcal{F}(n'; A', B', C', D') \quad \Rightarrow \quad |n| \le |n'|.$$

An LFT is called **trim** *if, with the partitions*

$$A = \begin{bmatrix} A_{1-} \\ \vdots \\ A_{r-} \end{bmatrix}, \quad A_{i-} \in \mathbb{F}^{n_i \times n}, \quad B = \begin{bmatrix} B_1 \\ \vdots \\ B_r \end{bmatrix}, \quad B_i \in \mathbb{F}^{n_i \times m}$$

we have that for $1 \le i \le r$, *each matrix* $\begin{bmatrix} A_{i-} & B_i \end{bmatrix}$ *has full row rank. Dually,* $(n; A, B, C, D)$ *is said to be* **co-trim** *iff, with*

$$A = \begin{bmatrix} A_{-1} & \cdots & A_{-r} \end{bmatrix}, \quad A_{-i} \in \mathbb{F}^{n \times n_i},$$

$$C = \begin{bmatrix} C_1 & \cdots & C_r \end{bmatrix}, \quad C_i \in \mathbb{F}^{g \times n_i}$$

each $\begin{bmatrix} A_{-i} \\ C_i \end{bmatrix}$ *has full column rank.*

Theorem 26 *A minimal LFT is both trim and co-trim. Conversely, trimness and co-trimness are not sufficient for minimality.*

Proof. A procedure for reducing the state dimension of non-trim LFTs is given below. The result for co-trimness is dual. An example of an LFT that is both trim and co-trim, but not state-minimal will be given in Section 6.2.1. □

An obvious necessary condition for minimality of $(n; A, B, C, D)$ is that the matrix pairs

$$\left(A_{ii}, \begin{bmatrix} A_{i1} & \cdots & A_{i,i-1} & A_{i,i+1} & \cdots & A_{ir} & B_i \end{bmatrix} \right), \quad 1 \le i \le r, \quad (6.7)$$

are controllable. Otherwise, the LFT can be reduced via a Kalman decomposition. This should be compared with Lemma 5.2.1. The minimality discussed there is based on a different equivalence concept, namely: $(n; A, B, C, D) \sim (n'; A', B', C', D')$ iff they are ON representations of the same behavior. This corresponds to Kuijper's notion of "strong external equivalence" [36]. Note that this equivalence does *not* imply that

$$\mathcal{F} := \mathcal{F}(n; A, B, C, D) = \mathcal{F}(n'; A', B', C', D') =: \mathcal{F}'$$

which is the relevant equivalence definition in the present context. (However, it *does* imply that the rational matrices $\mathcal{F}$ and $\mathcal{F}'$ have the same row span; this amounts to "weak external equivalence" [36].)

Via the Hautus test, one can see that trimness, that is

$$\text{rank} \begin{bmatrix} A_{i-} & B_i \end{bmatrix} = n_i \quad \text{for all } i$$

is in turn necessary for controllability of (6.7).

If $\Delta - A$ is invertible over the Laurent polynomial ring, the matrices A_{ii} are nilpotent according to Theorem 22. This implies that controllability of (6.7) is even equivalent to trimness of $(n; A, B, C, D)$. A dual argument applies to co-trimness and observability.

For $1 \leq i \leq r$, let

$$M_i := [\ A_{i-}\ \ B_i\ ] = [\ A_{i1}\ \ \cdots\ \ A_{ir}\ \ B_i\ ], \qquad \rho_i := \text{rank}(M_i) \quad (6.8)$$

and define $\rho := (\rho_1, \ldots, \rho_r)$, $\rho := |\rho|$. We will construct matrices $\begin{bmatrix} \tilde{A} & \tilde{B} \\ \tilde{C} & \tilde{D} \end{bmatrix} \in$ $\mathbb{F}^{(\rho+g) \times (\rho+q)}$ such that

$$\mathcal{F}(n; A, B, C, D) = \mathcal{F}(\rho; \tilde{A}, \tilde{B}, \tilde{C}, \tilde{D}). \qquad (6.9)$$

The new LFT $(\rho; A, B, C, D)$ is not necessarily trim, but by successively re-applying the reduction technique above, one obtains a sequence of representations of decreasing state-dimension which eventually must become stationary.

Reduction technique: Compute a column-echelon form (cef) of each M_i, that is (up to a permutation of rows) a matrix of the form

$$\begin{bmatrix} I_{\rho_i} & 0 \\ X_i & 0 \end{bmatrix}$$

that can be obtained from M_i by elementary column operations. In other words,

$$\text{cef}(M_i) = M_i P_i = \Pi_i \begin{bmatrix} I_{\rho_i} & 0 \\ X_i & 0 \end{bmatrix}$$

with a permutation matrix Π_i and an invertible matrix P_i. Define

$$T_i := [\ I_{\rho_i}\ \ 0\ ] \Pi^{-1} \in \mathbb{F}^{\rho_i \times n_i} \quad \text{and} \quad \hat{T}_i := \Pi_i \begin{bmatrix} I_{\rho_i} \\ X_i \end{bmatrix} \in \mathbb{F}^{n_i \times \rho_i}.$$

Then

$$\hat{T}_i T_i M_i = M_i. \qquad (6.10)$$

Note that although T_i and $\hat{T}_i$ are defined in terms of the permutation Π_i, they can be read off directly from the column-echelon form (e.g., $\hat{T}_i$ consists of the first ρ_i columns of cef(M_i)). Finally, define $T := \text{diag}(T_1, \ldots, T_r)$, $\hat{T} = \text{diag}(\hat{T}_1, \ldots, \hat{T}_r)$ and

$$\begin{aligned} \tilde{A} &:= TA\hat{T} \\ \tilde{B} &:= TB \\ \tilde{C} &:= C\hat{T} \\ \tilde{D} &:= D. \end{aligned}$$

Lemma 6.2.1. *The matrices* $\tilde{A}, \tilde{B}, \tilde{C}, \tilde{D}$ *defined above satisfy* (6.9).

Proof. Let $\Delta = \text{diag}(\delta_1 I_{n_1}, \dots, \delta_r I_{n_r})$ and $\tilde{\Delta} = T\Delta\hat{T} = \text{diag}(\delta_1 I_{\rho_1}, \dots, \delta_r I_{\rho_r})$. Note that

$$\hat{T}\tilde{\Delta} = \Delta\hat{T},$$

hence

$$\hat{T}(\tilde{\Delta} - TA\hat{T}) = (\Delta - \hat{T}TA)\hat{T}$$

or

$$(\Delta - \hat{T}TA)^{-1}\hat{T} = \hat{T}(\tilde{\Delta} - TA\hat{T})^{-1}.$$

By (6.10),

$$\hat{T}T\,[\,A \quad B\,] = [\,A \quad B\,],$$

and thus

$$
\begin{aligned}
C(\Delta - A)^{-1}B &= C(\Delta - \hat{T}TA)^{-1}\hat{T}TB \\
&= C\hat{T}(\tilde{\Delta} - TA\hat{T})^{-1}TB \\
&= \tilde{C}(\tilde{\Delta} - \tilde{A})^{-1}\tilde{B},
\end{aligned}
$$

which proves $\mathcal{F}(n; A, B, C, D) = \mathcal{F}(\rho; \tilde{A}, \tilde{B}, \tilde{C}, \tilde{D})$ as required. $\qquad\square$

The effectiveness of the reduction algorithm is illustrated by the following two examples.

Example 6.2.1. Consider $R(\delta_1, \delta_2) = a\delta_1\delta_2^2 + b\delta_1 + c\delta_2^2 + d$, where $a, b, c, d \in \mathbb{R}$ and $a \neq 0$. A direct representation is given by $R = \mathcal{F}((2,2); A, B, C, D)$ with

$$
\left[\begin{array}{c|c} A & B \\ \hline C & D \end{array}\right] =
\left[\begin{array}{cccc|c}
0 & 0 & 0 & 0 & a \\
0 & 0 & 0 & 0 & b \\
0 & 0 & 0 & 1 & 0 \\
1 & 0 & 0 & 0 & c \\
\hline
0 & 1 & 1 & 0 & d
\end{array}\right].
$$

We have $\rho = (1, 2)$ and

$$
T = \begin{bmatrix} 1 & 0 & 0 & 0 \\ 0 & 0 & 1 & 0 \\ 0 & 0 & 0 & 1 \end{bmatrix}, \qquad
\hat{T} = \begin{bmatrix} 1 & 0 & 0 \\ b/a & 0 & 0 \\ 0 & 1 & 0 \\ 0 & 0 & 1 \end{bmatrix}.
$$

Thus the reduction step yields $R = \mathcal{F}((1,2); \tilde{A}, \tilde{B}, \tilde{C}, \tilde{D})$ with

$$\left[\begin{array}{c|c} \tilde{A} & \tilde{B} \\ \hline C & D \end{array}\right] = \left[\begin{array}{ccc|c} 0 & 0 & 0 & a \\ \hline 0 & 0 & 1 & 0 \\ 1 & 0 & 0 & c \\ \hline b/a & 1 & 0 & d \end{array}\right]$$

which is in trim form. This example is considered by Cockburn and Morton [17] with $a = 15$, $b = 6$, $c = 5$, $d = 2$. Despite appearances, the possibility of reducing the order to 3 is *not* due to the fact that R factors as $R = (1 + 3\delta_1)(2 + 5\delta_2^2)$.

Example 6.2.2. In [16], Cheng and DeMoor give an LFT representation of

$$R(\delta_1, \delta_2) = \left[\begin{array}{cc} 1 + \dfrac{\delta_2}{1 + 2\delta_1} & \dfrac{5\delta_1\delta_2}{1 + 6\delta_1\delta_2} \\ \dfrac{5\delta_1\delta_2}{1 + 6\delta_1\delta_2} & 1 + \dfrac{3\delta_1}{1 + 4\delta_2} \end{array}\right]$$

of order 12 with $n = (6, 6)$. Reducing the given representation by the method described above, we obtain an LFT of order 8, namely

$$R = \mathcal{F}((4, 4); A, B, C, D)$$

with

$$\left[\begin{array}{c|c} A & B \\ \hline C & D \end{array}\right] = \left[\begin{array}{cccc|cccc|cc} -2 & 0 & 0 & 0 & 1 & 0 & 0 & 0 & 0 & 0 \\ 0 & 0 & 0 & 0 & 0 & 0 & 0 & 0 & 0 & 1 \\ 0 & 0 & 0 & 0 & 0 & 0 & -6 & 0 & 1 & 0 \\ 0 & 0 & 0 & 0 & 0 & 0 & 0 & -6 & 0 & 1 \\ \hline 0 & 0 & 0 & 0 & 0 & 0 & 0 & 0 & 1 & 0 \\ 0 & 3 & 0 & 0 & 0 & -4 & 0 & 0 & 0 & 0 \\ 0 & 0 & 1 & 0 & 0 & 0 & 0 & 0 & 0 & 0 \\ 0 & 0 & 0 & 1 & 0 & 0 & 0 & 0 & 0 & 0 \\ \hline -2 & 0 & 0 & 0 & 1 & 0 & 0 & 5 & 1 & 0 \\ 0 & 3 & 0 & 0 & 0 & -4 & 5 & 0 & 0 & 1 \end{array}\right]$$

that is, the number of states can be reduced by one third.

6.2.1 Further Transformations

Sugie [72] suggests another transformation of the representing matrices of an LFT, by swapping certain rows and columns that belong to different blocks. Although such a permutation does not change the state space dimension, it may nevertheless yield a further possibility of reduction, as it may destroy the original representation's property of being both trim and co-trim (and then the reduction technique of the previous section can be re-applied). In other words, the given representation is first reduced to trim and co-trim form, say

$(n; A, B, C, D)$. Then one searches for an appropriate pair of indices (ξ, η) such that

$$\mathcal{F}(n; A, B, C, D) = \mathcal{F}(n; \Pi_{\xi\eta} A \Pi_{\xi\eta}, \Pi_{\xi\eta} B, C\Pi_{\xi\eta}, D) \qquad (6.11)$$

and the representation on the right hand side is either not trim or not co-trim, that is, it is again reducible according to Section 6.2. Here, $\Pi_{\xi\eta}$ denotes the permutation matrix that interchanges the ξ-th and η-th row (or column, respectively). The crucial point is that the ξ-th and η-th row of A and B belong to different blocks M_i with respect to (6.8). Otherwise, $\Pi_{\xi\eta}$ is just a transformation that respects the block partition of Δ, in particular, it does not affect trimness. Note that the Δ-matrices that are tacitly involved in (6.11) are the same, that is, with $\Delta = \text{diag}(\delta_1 I_{n_1}, \dots, \delta_r I_{n_r})$, the claim of (6.11) is

$$C(\Delta - A)^{-1}B \equiv C\Pi_{\xi\eta}(\Delta - \Pi_{\xi\eta}A\Pi_{\xi\eta})^{-1}\Pi_{\xi\eta}B \qquad (6.12)$$

under suitable conditions on the index pair (ξ, η) and the matrices A, B, C. If ξ, η belong to the same block, we have $\Pi_{\xi\eta}\Delta\Pi_{\xi\eta} = \Delta$ and then (6.12) is trivially true.

Theorem 27 *[72] Let $(n; A, B, C, D)$ be given, and let $n = |n|$. Suppose that $1 \le \xi, \eta \le n$ are such that*

$$A_{\xi j} = 0 \quad \text{for all } j \ne \eta \qquad \text{and} \qquad A_{i\eta} = 0 \quad \text{for all } i \ne \xi$$

and

$$B_{\xi j} = 0 \quad \text{for all } j \qquad \text{and} \qquad C_{i\eta} = 0 \quad \text{for all } i.$$

Then equality (6.12) holds.

Proof. We use the feedback interpretation of the LFT. Let

$$\begin{bmatrix} z \\ y \end{bmatrix} = \begin{bmatrix} A & B \\ C & D \end{bmatrix} \begin{bmatrix} w \\ u \end{bmatrix}, \quad w = \Delta z.$$

Then $y = (C\Delta(I - A\Delta)^{-1}B + D)u$. By the assumption on A and B, we have $z_\xi = A_{\xi\eta}w_\eta$. We define vectors z_1 and w_1 such that with $\Pi := \Pi_{\xi\eta}$,

$$\begin{bmatrix} z_1 \\ y \end{bmatrix} = \begin{bmatrix} \Pi A \Pi & \Pi B \\ C\Pi & D \end{bmatrix} \begin{bmatrix} w_1 \\ u \end{bmatrix}, \quad w_1 = \Delta z_1,$$

i.e., $y = (C\Pi\Delta(I - \Pi A \Pi \Delta)^{-1}\Pi B + D)u$, which yields the desired result as u is arbitrary. Define

$$(z_1)_i = \begin{cases} z_i & \text{if } i \ne \xi, \eta \\ z_\eta & \text{if } i = \xi \\ \Delta_{\xi\xi} A_{\xi\eta} z_\eta & \text{if } i = \eta \end{cases}$$

and set

$$(w_1)_i := \Delta_{ii}(z_1)_i = \begin{cases} w_i & \text{if } i \neq \xi, \eta \\ \Delta_{\xi\xi} z_\eta & \text{if } i = \xi \\ w_\xi & \text{if } i = \eta, \end{cases}$$

where the last equality follows from $\Delta_{\eta\eta} \Delta_{\xi\xi} A_{\xi\eta} z_\eta = \Delta_{\xi\xi} A_{\xi\eta} w_\eta = \Delta_{\xi\xi} z_\xi = w_\xi$. Then $(w - \Pi w_1)_i = 0$ for all $i \neq \eta$ and hence $C(w - \Pi w_1) = 0$ due to the assumption on C. As $\Pi^{-1} = \Pi$, it remains to be shown that

$$\Pi z_1 = A \Pi w_1 + Bu.$$

In case that $i \neq \xi$, we have $(\Pi z_1)_i = z_i$ and $(A \Pi w_1)_i = (Aw)_i$. Finally,

$$(\Pi z_1)_\xi = (z_1)_\eta = \Delta_{\xi\xi} A_{\xi\eta} z_\eta = A_{\xi\eta}(w_1)_\xi = A_{\xi\eta}(\Pi w_1)_\eta = (A \Pi w_1)_\xi$$

concludes the proof. $\qquad\qquad\qquad\qquad\qquad\qquad\qquad\qquad\qquad\qquad$ □

Example 6.2.3. Consider the LFT with $n = (3, 1)$ given by

$$\left[\begin{array}{cccc|cc} 0 & 0 & 0 & 0 & 1 & 0 \\ 0 & 0 & 0 & 0 & 0 & 1 \\ 1 & 0 & 0 & 0 & 0 & 0 \\ \hline 0 & 1 & 0 & 0 & 0 & 0 \\ \hline 0 & 0 & 1 & 1 & 0 & 0 \end{array}\right].$$

It is both trim and co-trim, but $(\xi, \eta) = (4, 2)$ is a pair according to Theorem 27. Swapping yields

$$\left[\begin{array}{cccc|cc} 0 & 0 & 0 & 0 & 1 & 0 \\ 0 & 0 & 0 & 1 & 0 & 0 \\ 1 & 0 & 0 & 0 & 0 & 0 \\ \hline 0 & 0 & 0 & 0 & 0 & 1 \\ \hline 0 & 1 & 1 & 0 & 0 & 0 \end{array}\right],$$

which is not co-trim. This implies that the LFT can be reduced to $\rho = (2, 1)$. The result is

$$\left[\begin{array}{cc|cc} 0 & 0 & 0 & 1 & 0 \\ 1 & 0 & 1 & 0 & 0 \\ \hline 0 & 0 & 0 & 0 & 1 \\ \hline 0 & 1 & 0 & 0 & 0 \end{array}\right].$$

This example also demonstrates that an LFT that is trim and co-trim need not be minimal.

6.3 Balancing

Let $\mathbb{K}$ denote either the field of real or the field of complex numbers, and let A, P in $\mathbb{K}^{n \times n}$ be given. As usual, $\rho(A)$ denotes the spectral radius of A, and $\overline{\sigma}(A) = \|A\|_2$ is its largest singular value. The notation $P > 0$ ($P \geq 0$) indicates that P is Hermitian and positive definite (semi-definite). The set of non-singular $n \times n$ matrices over $\mathbb{K}$ is denoted by $Gl_n(\mathbb{K})$.

The equivalence of the following assertions is well-known from 1D systems theory.

1. A is discrete-time asymptotically stable, that is, $\rho(A) < 1$;
2. There exists an invertible matrix T such that $\overline{\sigma}(T^{-1}AT) < 1$;
3. There exists $P > 0$ such that $APA^* - P < 0$;
4. For every $Q \geq 0$, there exists $P > 0$ such that $APA^* - P + Q < 0$.

In view of Condition 2 above, we recall the well-known fact that similarity transforms

$$(A, B, C) \mapsto (T^{-1}AT, T^{-1}B, CT)$$

do not change the transfer function $H(s) = C(sI - A)^{-1}B$ of a linear 1D system. For LFT systems

$$\mathcal{F}(\Delta; A, B, C) := C(\Delta^{-1} - A)^{-1}B = C\Delta(I - A\Delta)^{-1}B$$

however, we only have

$$\mathcal{F}(\Delta; A, B, C) = \mathcal{F}(T^{-1}\Delta T; T^{-1}AT, T^{-1}B, CT)$$

since $T^{-1}\Delta T \neq \Delta$ in general (unlike the 1D case, where $\Delta = s^{-1}I$). Hence the admissible similarity transforms $T \in Gl_n(\mathbb{K})$ that leave the LFT invariant are those for which $\Delta T = T\Delta$. Let $\mathcal{C}$ denote the centralizer of Δ in $Gl_n(\mathbb{K})$, that is, the set of all non-singular matrices that commute with Δ. It consists of the matrices that share the block structure of $\Delta = \text{diag}(\delta_1 I_{n_1}, \ldots, \delta_r I_{n_r})$, i.e.,

$$\mathcal{C} = \{T \in Gl_n(\mathbb{K}), \Delta T = T\Delta\} = \{T = \text{diag}(T_1, \ldots, T_r), T_i \in Gl_{n_i}(\mathbb{K})\}.$$

This observation gives rise to the following definition of stability in the LFT setting:

Definition 40 *Let* $n \in \mathbb{N}^r$, $n = |n|$, *and* $A \in \mathbb{K}^{n \times n}$ *be given. The matrix* A *is said to be* n*-stable (i.e., stable with respect to the partition induced by* n, *or "Q-stable" in the nomenclature of [77]) if there exists an invertible* n*-block matrix* T, *that is,*

$$T = \text{diag}(T_1, \ldots, T_r) \quad \text{with } T_i \in Gl_{n_i}(\mathbb{K}) \text{ for } 1 \leq i \leq r,$$

such that $\overline{\sigma}(T^{-1}AT) < 1$.

An n-stable matrix is discrete-time asymptotically stable. For Hermitian matrices $A = A^*$, we have $\bar{\sigma}(A) = \rho(A)$, and hence the converse is also true. The set of non-negative matrices constitutes another interesting class of matrices for which $\rho(A) < 1$ already implies n-stability of A (for arbitrary block structures n).

Definition 41 *A real matrix is said to be* **positive (non-negative)** *if all its entries are positive (non-negative).*

Stoer and Witzgall [71] considered the problem of minimizing the norm of a positive matrix by scaling. They showed that, for a positive matrix A,

$$\min_{D \in D} \|DAD^{-1}\| = \rho(A),$$

where D denotes the set of non-singular diagonal matrices. This result holds for a class of matrix norms including the Hölder norms $\| \cdot \|_p$ for $1 \leq p \leq \infty$. Here we consider the case $p = 2$. The minimizing D is unique (up to a non-zero factor). Indeed, if $\pi(A) = \rho(A)$ denotes the Perron eigenvalue [27, Chapter 13] of A, let x and y be left and right Perron eigenvectors, *i.e.*,

$$x^T A = \pi(A)x^T \quad \text{and} \quad Ay = \pi(A)y,$$

then [66]

$$D = \text{diag}\left(\frac{\sqrt{x_1}}{\sqrt{y_1}}, \ldots, \frac{\sqrt{x_n}}{\sqrt{y_n}}\right)$$

minimizes $\|DAD^{-1}\|_2$.

Any non-negative matrix can be written as the limit of a sequence of positive matrices, and hence a continuity argument implies

$$\inf_{D \in D} \|DAD^{-1}\| = \rho(A) \tag{6.13}$$

for non-negative matrices A. Note that the infimum may not be achieved, as can be seen from the following example.

Example 6.3.1. Consider

$$A = \begin{bmatrix} 0 & 1 \\ 0 & 0 \end{bmatrix}$$

with $\rho(A) = 0$ and $\|A\|_2 = 1$. Let $D = \text{diag}(d_1, d_2) \in D$ be an arbitrary scaling matrix, then

$$DAD^{-1} = \begin{bmatrix} 0 & \frac{d_1}{d_2} \\ 0 & 0 \end{bmatrix}$$

and $\|DAD^{-1}\|_2 = \left|\frac{d_1}{d_2}\right| \neq 0$ for all $D \in D$. However, $\|DAD^{-1}\|_2$ can obviously be made arbitrarily small by choice of d_1, d_2.

Still, we conclude from (6.13) that for non-negative matrices A with $\rho(A) < 1$, we can find a $D \in \mathbf{D}$ such that

$$\|DAD^{-1}\|_2 = \overline{\sigma}(DAD^{-1}) < 1,$$

thus implying n-stability for arbitrary block structures n.

Equation (6.13) does not hold in the general case of matrices with arbitrary sign patterns. In fact, the value of $\inf_D \|DAD^{-1}\|_2$ can be arbitrarily far away from $\rho(A)$.

Example 6.3.2. Consider

$$A = \frac{\beta}{2} \begin{bmatrix} 1 & -1 \\ 1 & -1 \end{bmatrix}$$

with $\rho(A) = 0$ and $\|A\|_2 = \beta > 0$. The scaled versions of A can be parametrized as

$$A(\alpha) = \frac{\beta}{2} \begin{bmatrix} 1 & -\alpha \\ \alpha^{-1} & -1 \end{bmatrix}, \quad \text{where } \alpha \neq 0.$$

As $\|A(\alpha)\|_2 = \frac{\beta}{2}\sqrt{2 + \alpha^2 + \alpha^{-2}}$ takes its minimum at $\alpha = 1$, it turns out that A is already optimally scaled, and $\inf_D \|DAD^{-1}\|_2 = \beta$, whereas $\rho(A) = 0$.

In the general case (A not necessarily Hermitian or non-negative), we can compute

$$\mu(A) := \inf_{D \in \mathbf{D}} \|DAD^{-1}\|_2 \tag{6.14}$$

only approximatively. In fact, the MATLAB μ-Analysis and Synthesis Toolbox does precisely this job, it yields an estimate for $\mu(A)$ as well as a diagonal scaling matrix D for which this estimate is achieved.

For an arbitrary block structure n, we are interested in

$$\mu(A, n) = \inf\{\|T^{-1}AT\|_2, T \text{ has } n\text{-block structure}\} \leq \mu(A).$$

The computation of $\mu(A, n)$ and hence the decision problem "Is A n-stable?" can be reformulated in terms of a **linear matrix inequality (LMI)** using the subsequent lemma, which generalizes the equivalences given at the beginning of this section.

Lemma 6.3.1. Let $n \in \mathbb{N}^r$, $n = |n|$, and $A \in \mathbb{K}^{n \times n}$ be given. The following are equivalent:

1. A is n-stable;
2. There exists an invertible n-block matrix T such that $\overline{\sigma}(T^{-1}AT) < 1$;
3. There exists an n-block matrix $P > 0$ such that $APA^* - P < 0$;

4. *For every $Q \geq 0$, there exists an n-block matrix $P > 0$ such that*

$$APA^* - P + Q < 0.$$

Proof. Conditions 1 and 2 are equivalent by definition, Condition 3 is an obvious special case of 4 ($Q = 0$). An equivalent formulation of Condition 2 is

$$(T^{-1}AT)(T^{-1}AT)^* - I < 0.$$

This shows the equivalence of 2 and 3 by putting $P = TT^*$ or $T = P^{\frac{1}{2}}$, respectively. Then the block structure of T implies the block structure of P and vice versa. Finally, suppose that there exists an n-block matrix $P > 0$ such that $APA^* - P =: Q_1 < 0$, and let $Q \geq 0$ be given. Choose

$$\lambda > \rho(QQ_1^{-1}) = \max_{x \neq 0} \frac{x^* Q x}{x^* (-Q_1) x} \tag{6.15}$$

and define $P_1 := \lambda P > 0$. Then $AP_1A^* - P_1 + Q = \lambda Q_1 + Q < 0$ by the choice of λ, and P_1 inherits the n-block structure of P. □

Corollary 18 *Let A be non-negative and let $\rho(A) < 1$. Then there exists, for any $Q \geq 0$, a diagonal matrix $D > 0$ such that*

$$ADA^* - D + Q < 0.$$

Using the above lemma, we can reformulate the decision problem "Is A n-stable?" as follows: Let $P_1, \ldots, P_N \in \mathbb{K}^{n \times n}$ be an $\mathbb{R}$-basis of the space of Hermitian n-block matrices. For instance, for a 2×2 block and $\mathbb{K} = \mathbb{R}$, we might take

$$\begin{bmatrix} 1 & 0 \\ 0 & 0 \end{bmatrix}, \begin{bmatrix} 0 & 0 \\ 0 & 1 \end{bmatrix}, \begin{bmatrix} 0 & 1 \\ 1 & 0 \end{bmatrix},$$

and if $\mathbb{K} = \mathbb{C}$, we admit the additional generator

$$\begin{bmatrix} 0 & i \\ -i & 0 \end{bmatrix}.$$

In general, $N = \frac{1}{2} \sum_{i=1}^{r} n_i(n_i + 1)$ for $\mathbb{K} = \mathbb{R}$, and $N = \sum_{i=1}^{r} n_i^2$ for $\mathbb{K} = \mathbb{C}$. Define

$$F_i = \begin{bmatrix} P_i & 0 \\ 0 & P_i - AP_iA^* \end{bmatrix}$$

and consider the following standard form of an LMI [5]

$$F(x) = F(x_1, \ldots, x_N) = \sum_{i=1}^{N} x_i F_i > 0. \tag{6.16}$$

This problem is **feasible** if the convex set $\{x \in \mathbb{R}^N, F(x) > 0\}$ is non-empty. If x_0 is a solution (that is, if $F(x_0) > 0$), then $P := \sum_i x_i P_i > 0$ has n-block structure and $APA^* - P < 0$ as desired. Infeasibility of (6.16) is equivalent to the existence of $G \geq 0$, $G \neq 0$ such that [5]

$$\text{trace}(GF_i) = 0 \quad \text{for } i = 1, \ldots, N.$$

Definition 42 *Let $A \in \mathbb{K}^{n \times n}$, $B \in \mathbb{K}^{n \times q}$, and $C \in \mathbb{K}^{g \times n}$ be given. The matrix triple (A, B, C) is said to be **balanced** if there exists a real diagonal matrix $\Sigma > 0$ such that*

$$A\Sigma A^* - \Sigma + BB^* < 0 \quad \text{and} \quad A^*\Sigma A - \Sigma + C^*C < 0.$$

It is a common fact that if A is discrete-time asymptotically stable, every matrix triple (A, B, C) possesses an equivalent triple $(T^{-1}AT, T^{-1}B, CT)$ that is balanced. Indeed, let $P > 0$ and $Q > 0$ be such that $APA^* - P + BB^* < 0$ and $A^*QA - Q + C^*C < 0$, and define $R := P^{\frac{1}{2}}QP^{\frac{1}{2}} > 0$. Let U be the unitary matrix that transforms R into diagonal form, that is, $U^*RU = \Lambda = \text{diag}(\lambda_1, \ldots, \lambda_n)$. Being the eigenvalues of the Hermitian matrix R, the diagonal entries of Λ have to be real. Then $T = P^{\frac{1}{2}}U\Lambda^{-\frac{1}{4}}$ has the desired property with $\Sigma = \Lambda^{\frac{1}{2}}$. Note that neither T nor Σ are unique. The following result on n-stable matrices is an immediate consequence:

Lemma 6.3.2. *Let A be n-stable. Then there exists an invertible n-block matrix T such that $(T^{-1}AT, T^{-1}B, CT)$ is balanced.*

Corollary 19 *Let A be non-negative and let $\rho(A) < 1$. Then there exits a non-singular diagonal matrix D such that (DAD^{-1}, DB, CD^{-1}) is balanced.*

6.3.1 Balanced Truncation

Let (A, B, C) be balanced with Gramian Σ. Let $n \in \mathbb{N}^r$ be a given block structure with $|n| = n$, and let $\Sigma = \text{diag}(\Sigma_1, \ldots, \Sigma_r)$ be the corresponding partition of Σ with

$$\mathbb{R}^{n_i \times n_i} \ni \Sigma_i = \text{diag}(\sigma_{i1} I_{n_{i1}}, \ldots, \sigma_{il(i)} I_{n_{il(i)}})$$

where $\sigma_{i1} > \ldots > \sigma_{il(i)}$. The integers n_{ij} with $\sum_{j=1}^{l(i)} n_{ij} = n_i$ are defined by the number of diagonal entries of Σ_i equal to σ_{ij}. Now partition each $\Sigma_i = \text{diag}(\Sigma_i^{(1)}, \Sigma_i^{(2)})$ such that

$$\Sigma_i^{(1)} = \text{diag}(\sigma_{i1} I_{n_{i1}}, \ldots, \sigma_{ik(i)} I_{n_{ik(i)}}),$$
$$\Sigma_i^{(2)} = \text{diag}(\sigma_{ik(i)+1} I_{n_{ik(i)+1}}, \ldots, \sigma_{il(i)} I_{n_{il(i)}})$$

for some $0 \leq k(i) \leq l(i)$. Let ν_i denote the size of $\Sigma_i^{(1)}$ and $\nu = (\nu_1, \ldots, \nu_r)$.

Consider the corresponding partitions of A, B, and C:

$$A = \begin{bmatrix} A_{11} & \cdots & A_{1r} \\ \vdots & & \vdots \\ A_{r1} & \cdots & A_{rr} \end{bmatrix}, \ \mathbb{K}^{n_i \times n_j} \ni A_{ij} = \begin{bmatrix} A_{ij}^{(1)} & A_{ij}^{(2)} \\ A_{ij}^{(3)} & A_{ij}^{(4)} \end{bmatrix}, \ A_{ij}^{(1)} \in \mathbb{K}^{\nu_i \times \nu_j},$$

$$B = \begin{bmatrix} B_1 \\ \vdots \\ B_r \end{bmatrix}, \ \mathbb{K}^{n_i \times q} \ni B_i = \begin{bmatrix} B_i^{(1)} \\ B_i^{(2)} \end{bmatrix}, \ B_i^{(1)} \in \mathbb{K}^{\nu_i \times q},$$

$$C = \begin{bmatrix} C_1 & \cdots & C_r \end{bmatrix}, \ \mathbb{K}^{g \times n_i} \ni C_i = \begin{bmatrix} C_i^{(1)} & C_i^{(2)} \end{bmatrix}, \ C_i^{(1)} \in \mathbb{K}^{g \times \nu_i}.$$

Finally, let $\hat{\Delta} = \text{diag}(\delta_1 I_{\nu_1}, \ldots, \delta_r I_{\nu_r})$ and

$$\hat{A} = \begin{bmatrix} A_{11}^{(1)} & \cdots & A_{1r}^{(1)} \\ \vdots & & \vdots \\ A_{r1}^{(1)} & \cdots & A_{rr}^{(1)} \end{bmatrix}, \quad \hat{B} = \begin{bmatrix} B_1^{(1)} \\ \vdots \\ B_r^{(1)} \end{bmatrix}, \quad \hat{C} = \begin{bmatrix} C_1^{(1)} & \cdots & C_r^{(1)} \end{bmatrix}.$$

Then the balanced truncation

$$\hat{Q} := \mathcal{F}(\hat{\Delta}; \hat{A}, \hat{B}, \hat{C})$$

is an approximation of $Q = \mathcal{F}(\Delta; A, B, C)$ in the sense that $\hat{A}$ is ν-stable and [77]

$$\|Q - \hat{Q}\|_\infty \leq 2 \sum_{i=1}^r \sum_{j=k(i)+1}^{l(i)} \sigma_{ij},$$

where

$$\|Q\|_\infty = \sup_{|\delta_i| \leq 1} \|Q(\delta_1, \ldots, \delta_r)\|_2.$$

Note that the existence of $\|Q\|_\infty$, that is, the boundedness of $\|Q(\delta_1, \ldots, \delta_r)\|_2$ on the closed multi-disk

$$\mathbb{D}^r = \{\delta = (\delta_1, \ldots, \delta_r) \in \mathbb{C}^r, |\delta_i| \leq 1 \text{ for } 1 \leq i \leq r\}$$

is a consequence of the n-stability of A. This is proven in the following theorem.

Theorem 28 *Let A be n-stable. Then for all $\delta \in \mathbb{D}^r$, we have $\det(I - A\Delta(\delta)) \neq 0$, where $\Delta(\delta) = (\delta_1 I_{n_1}, \ldots, \delta_r I_{n_r}) \in \mathbb{C}^{n \times n}$.*

Proof. First consider the case when $\Delta(\delta)$ is non-singular and suppose that

$$\det(\Delta(\delta)^{-1} - A) = 0.$$

We need to prove that $\delta \notin \mathbb{D}^r$. There exists a complex vector $z \neq 0$ such that $z^*\Delta(\delta)^{-1} = z^*A$. On the other hand, there exists an n-block matrix $P > 0$ such that $APA^* - P < 0$. Hence $z^*(APA^* - P)z = z^*(\Delta(\delta)^{-1}P\Delta(\delta)^{-*} - P)z < 0$. Now

$$\Delta(\delta)^{-1}P\Delta(\delta)^{-*} = \begin{bmatrix} \frac{1}{|\delta_1|^2}P_1 & & \\ & \ddots & \\ & & \frac{1}{|\delta_r|^2}P_r \end{bmatrix}$$

and we obtain, with the n-partitioning of $z = (z_1^T, \ldots, z_r^T)^T$,

$$\sum_{i=1}^{r} \left(\frac{1}{|\delta_i|^2} - 1 \right) z_i^* P_i z_i < 0.$$

Since $z_i^* P_i z_i \geq 0$, we must have $|\delta_i| > 1$ for at least one i.

Now assume that $\Delta(\delta)$ is singular. The case $\Delta(\delta) = 0$ is trivial, so without loss of generality, we have

$$\Delta(\delta) = \begin{bmatrix} \Delta_1(\delta_1) & 0 \\ 0 & 0 \end{bmatrix}$$

with a non-singular matrix $\Delta_1(\delta_1) = \mathrm{diag}(\delta_1 I_{n_1}, \ldots, \delta_\rho I_{n_\rho})$. Let

$$A = \begin{bmatrix} A_1 & A_2 \\ A_3 & A_4 \end{bmatrix}$$

be the corresponding partition of A and suppose that $\det(I - A\Delta(\delta)) = 0$. Then $\det(I - A_1\Delta_1(\delta_1)) = 0$ and we are in the situation discussed above, if we can show that A_1 is ν-stable, where $\nu = (n_1, \ldots, n_\rho)$. We have $APA^* - P < 0$ for some $P = \mathrm{diag}(P_1, P_2) > 0$. Plugging in the conforming partitions of A and P,

$$\begin{bmatrix} A_1P_1A_1^* + A_2P_2A_2^* - P_1 & A_1P_1A_3^* + A_2P_2A_4^* \\ A_3P_1A_1^* + A_4P_2A_2^* & A_3P_1A_3^* + A_4P_2A_4^* - P_2 \end{bmatrix} < 0.$$

In particular, $A_1P_1A_1^* - P_1 \leq A_1P_1A_1^* + A_2P_2A_2^* - P_1 < 0$. $\qquad\square$

6.4 LFTs and Robust Stability

This section focuses on (1D) general linear constant differential systems in which the coefficients depend polynomially on several parameters. We will see that the system matrix can be written in terms of a linear fractional transformation (LFT), which is a representation that extracts the parametric uncertainty. The LFT form yields lower bounds for the robust stability radius of the system via μ-analysis tools. The method is applied to the linearized model of a transistor amplifier.

One of the major problems with the robustness analysis of linear electrical circuits is the complexity of the symbolic transfer functions in the presence of several uncertain network parameters. The only compact description of such a circuit is usually a general linear constant differential system in the frequency domain

$$
\begin{aligned}
T(s;p)\xi(s) &= U(s;p)u(s) \\
y(s) &= V(s;p)\xi(s) + W(s;p)u(s),
\end{aligned}
$$

where u, ξ and y denote the Laplace transforms of input, state and output, respectively. The system matrices T, U, V, W depend polynomially on s and on the parameter vector $p = (p_1, \ldots, p_r)$ which contains the network elements such as resistances, capacitances etc. The system equations are generated from the topological structure underlying the circuit (the associated network graph) and the current–voltage relations corresponding to the individual network elements. Chapter 7 gives a brief overview on how to set up these equations. The main goal here consists in investigating the induced rational transfer function

$$
G(s;p) = V(s;p)T(s;p)^{-1}U(s;p) + W(s;p)
$$

without directly computing it, *i.e.*, symbolic inversion of $T(s;p)$ is avoided. This is due to the fact that even for small circuits (see Section 6.4.2 for an example), the complexity of the symbolic expression makes an evaluation virtually impossible.

Chapter 6 provides a representation of $G(s;p)$ in terms of a linear fractional transformation, that is,

$$
G(s;p) = G_{22}(s) + G_{21}(s)\Delta(p)(I - G_{11}(s)\Delta(p))^{-1}G_{12}(s).
$$

The parameters p are thus "extracted," that is, they enter the system in a simple feedback form. The corresponding coefficient matrices $G_{ij}(s)$ are rational functions of s. Their determination involves only the inversion of $T(s;v)$, where $v \in \mathbb{R}^r$ is a given numerical design point, a vector of nominal parameter values.

The open right half of the complex plane will be denoted by $\mathbb{C}_+ = \{s \in \mathbb{C}, \text{Re}(s) > 0\}$, its closure by $\overline{\mathbb{C}}_+ = \{s \in \mathbb{C}, \text{Re}(s) \geq 0\}$. A rational matrix is said to be *stable* iff it has no poles in $\overline{\mathbb{C}}_+$.

Let $\mathbb{F}$ denote the field of meromorphic functions on $\mathbb{C}_+$. Consider a parametrized linear system in polynomial form [63]

$$T(s; \boldsymbol{p})\xi(s) \;=\; U(s; \boldsymbol{p})u(s) \tag{6.17}$$
$$y(s) \;=\; V(s; \boldsymbol{p})\xi(s) + W(s; \boldsymbol{p})u(s) \tag{6.18}$$

where $u \in \mathbb{F}^m$, $\xi \in \mathbb{F}^\nu$, $y \in \mathbb{F}^p$, and $\boldsymbol{p} = (p_1, \ldots, p_r)$ is a parameter vector. Assume that the entries of T, U, V, W depend on $\boldsymbol{p}$ polynomially, that is, e.g., for an entry T_{ij} of $T(s; \boldsymbol{p})$,

$$
\begin{aligned}
T_{ij} &= \sum_{n \in \mathbb{N}^r} t_{ij}(s; n) \boldsymbol{p}^n \\
&= \sum_{n_1, \ldots, n_r \in \mathbb{N}} t_{ij}(s; n_1, \ldots, n_r) p_1^{n_1} \cdots p_r^{n_r}
\end{aligned}
$$

with coefficients $t_{ij}(s; n) \in \mathbb{R}[s]$ and a finite support

$$\operatorname{supp}(T_{ij}) = \{ n \in \mathbb{N}^r, \ t_{ij}(s; n) \not\equiv 0 \}.$$

Thus, with $\mathcal{P} := \mathbb{R}[s][\boldsymbol{p}] = \mathbb{R}[s, \boldsymbol{p}]$, the polynomial ring over $\mathbb{R}$ in $r + 1$ indeterminates s and $p_1, \ldots, p_r$, the matrix T is in $\mathcal{P}^{\nu \times \nu}$, $U \in \mathcal{P}^{\nu \times m}$, $V \in \mathcal{P}^{p \times \nu}$, $W \in \mathcal{P}^{p \times m}$.

Let the nominal value of each parameter p_i be given by $v_i \in \mathbb{R}$, and let the uncertainty, scaled via $m_i \in \mathbb{R}$, be denoted by δ_i: One considers $p_i = v_i + m_i \delta_i$ ($m_i = 1$ corresponds to additive uncertainty, and $m_i = v_i$ models a multiplicative uncertainty). Define

$$\boldsymbol{v} = (v_1, \ldots, v_r) \quad \text{and} \quad \boldsymbol{\delta} = (\delta_1, \ldots, \delta_r).$$

By expanding the binomials $p_i{}^{n_i} = (v_i + m_i \delta_i)^{n_i}$, elements of $\mathcal{P} = \mathbb{R}[s][\boldsymbol{p}]$ can be interpreted as polynomials in $\mathcal{D} := \mathbb{R}[s][\boldsymbol{\delta}]$ in a natural way:

For $a = \sum_n a(n) \boldsymbol{p}^n$, let $N := \operatorname{supp}(a) \subset \mathbb{N}^r$ denote the support of a, then

$$
\begin{aligned}
a &= \sum_{n \in N} a(n) \prod_{i=1}^{r} (v_i + m_i \delta_i)^{n_i} \\
&= \sum_{n \in N} a(n) \prod_{i=1}^{r} \sum_{j_i = 0}^{n_i} \binom{n_i}{j_i} v_i^{n_i - j_i} m_i^{j_i} \delta_i^{j_i} \\
&= \sum_{j \in J} \sum_{n \in N} a(n) \prod_{i=1}^{r} \binom{n_i}{j_i} v_i^{n_i - j_i} m_i^{j_i} \delta_i^{j_i} \\
&= \sum_{j \in J} \tilde{a}(j) \delta^j,
\end{aligned}
$$

where

$$J := \{ j \in \mathbb{N}^r, \ \exists n \in N \text{ with } j_i \leq n_i \text{ for } i = 1, \ldots, r \}.$$

This shows that $a = \sum a(n)p^n = \sum \tilde{a}(j)\delta^j$ is a polynomial in the indeterminates $\delta_1, \ldots, \delta_r$, where the coefficients are related via

$$\tilde{a}(j) := \sum_{n \in N} a(n) \prod_{i=1}^{r} \binom{n_i}{j_i} v_i^{n_i - j_i} m_i^{j_i}.$$

In the following, it will be assumed that the entries of T, U, V, W are in $\mathcal{D}$, i.e., one considers the polynomial system

$$T(s; \delta)\xi(s) \quad = \quad U(s; \delta)u(s) \tag{6.19}$$
$$y(s) \quad = \quad V(s; \delta)\xi(s) + W(s; \delta)u(s) \tag{6.20}$$

with nominal parameter point $\delta = 0$.

The following will be assumed throughout this section: The nominal system is stable in the sense that $\det(T(s; 0))$ is a Hurwitz polynomial, i.e., it has no zeros in $\overline{\mathbb{C}}_+$. The notation

$$G = \left[\begin{array}{c|c} T & U \\ \hline -V & W \end{array} \right]$$

will be used for the rational transfer function given by the corresponding polynomial system, i.e.,

$$G = VT^{-1}U + W.$$

The problem of analyzing robust stability in the situation described above will be attacked in the following way:

1. Define the coefficient matrix

$$M(s; \delta) = \left[\begin{array}{cc} T(s; \delta) & U(s; \delta) \\ -V(s; \delta) & W(s; \delta) \end{array} \right] \in \mathcal{D}^{g \times q},$$

where $g = \nu + p$ and $q = \nu + m$.

2. Write $M(s; \delta)$ in terms of a linear fractional transformation (LFT)

$$M(s; \delta) \quad = \quad \mathcal{F}(\Delta; M(s)) \tag{6.21}$$
$$= \quad M_{22}(s) + M_{21}(s)\Delta(I - M_{11}(s)\Delta)^{-1}M_{12}(s),$$

where for some suitable non-negative integers $n_1, \ldots, n_r$ and $n := \sum n_i$,

$$\Delta = \operatorname{diag}(\delta_1 I_{n_1}, \ldots, \delta_r I_{n_r}) \quad \text{and}$$

$$M(s) = \left[\begin{array}{cc} M_{11}(s) & M_{12}(s) \\ M_{21}(s) & M_{22}(s) \end{array} \right] \in \mathbb{R}[s]^{(n+g) \times (n+q)}.$$

In this representation, $M_{22}(s) = M(s; 0)$ is the nominal coefficient matrix, i.e., writing $T_0(s) := T(s; 0)$ and so on, (6.21) takes the form

$$M(s; \delta) = \begin{bmatrix} T_0(s) & U_0(s) \\ -V_0(s) & W_0(s) \end{bmatrix} + \\ \begin{bmatrix} M_{21}^a(s) \\ M_{21}^b(s) \end{bmatrix} \Delta (I - M_{11}(s)\Delta)^{-1} \begin{bmatrix} M_{12}^a(s) & M_{12}^b(s) \end{bmatrix},$$ (6.22)

where

$$M_{21}(s) = \begin{bmatrix} M_{21}^a(s) \\ M_{21}^b(s) \end{bmatrix} \text{ and } M_{12}(s) = \begin{bmatrix} M_{12}^a(s) & M_{12}^b(s) \end{bmatrix}$$

are partitions that match the corresponding partitions of $M(s; \delta)$.
3. Define

$$G_{11}(s) := M_{11}(s) - M_{12}^a(s) T_0(s)^{-1} M_{21}^a(s) \in \mathbb{R}(s)^{n \times n}$$ (6.23)

and

$$\boldsymbol{\Delta} = \{\Delta = \mathrm{diag}(\delta_1 I_{n_1}, \ldots, \delta_r I_{n_r}), \delta_i \in \mathbb{C}\} \subset \mathbb{C}^{n \times n}.$$

For $s \in i\mathbb{R}$, determine $\mu_{\boldsymbol{\Delta}}(G_{11}(s))$, which is defined by

$$\mu_{\boldsymbol{\Delta}}(G_{11}(s)) := \max\{\|\Delta\|^{-1}, \det(I - G_{11}(s)\Delta) = 0, \Delta \in \boldsymbol{\Delta}\}$$

unless there is no $\Delta \in \boldsymbol{\Delta}$ that makes $I - G_{11}(s)\Delta$ singular, in which case $\mu_{\boldsymbol{\Delta}}(G_{11}(s)) := 0$. Finally, let

$$\rho := \sup_{\omega \in \mathbb{R}} \mu_{\boldsymbol{\Delta}}(G_{11}(i\omega)).$$

Theorem 29 *In the situation described above, the system given by (6.19), (6.20) is robustly stable for all parameter variations with $\max_i |\delta_i| < \rho^{-1}$.*

Proof. Plugging the expressions for $T(s; \delta)$ and $U(s; \delta)$ from (6.22) into (6.19), one obtains

$$\begin{aligned} T_0(s)\xi(s) &= U_0(s)u(s) + M_{21}^a(s)\Delta(I - M_{11}(s)\Delta)^{-1} \cdot \\ &\quad \cdot (M_{12}^b(s)u(s) - M_{12}^a(s)\xi(s)). \end{aligned}$$ (6.24)

Define

$$\begin{aligned} z(s) &:= (I - M_{11}(s)\Delta)^{-1}(M_{12}^b(s)u(s) - M_{12}^a(s)\xi(s)), \\ w(s) &:= \Delta z(s), \end{aligned}$$

then (6.24) reads

$$T_0(s)\xi(s) = M_{21}^a(s)w(s) + U_0(s)u(s).$$

Similarly, using the LFTs for $V(s; \delta)$ and $W(s; \delta)$ from (6.22) in (6.20),

$$
\begin{aligned}
y(s) &= V_0(s)\xi(s) + M_{21}^b\Delta(I - M_{11}(s)\Delta)^{-1} \cdot \\
&\quad \cdot (M_{12}^b(s)u(s) - M_{12}^a(s)\xi(s)) + W_0(s)u(s) \\
&= V_0(s)\xi(s) + M_{21}^b(s)w(s) + W_0(s)u(s).
\end{aligned}
$$

Summing up, one obtains the following relations: $w = \Delta z$, and

$$
\begin{aligned}
T_0\xi &= M_{21}^a w + U_0 u \\
z &= -M_{12}^a\xi + M_{11}w + M_{12}^b u \\
y &= V_0\xi + M_{21}^b w + W_0 u.
\end{aligned}
$$

The corresponding transfer function from $(w, u)^T$ to $(z, y)^T$ is

$$
G = \begin{bmatrix} -M_{12}^a \\ V_0 \end{bmatrix} T_0^{-1} \begin{bmatrix} M_{21}^a & U_0 \end{bmatrix} + \begin{bmatrix} M_{11} & M_{12}^b \\ M_{21}^b & W_0 \end{bmatrix},
$$

or,

$$
G = \begin{bmatrix} G_{11} & G_{12} \\ G_{21} & G_{22} \end{bmatrix} = \left[\begin{array}{c|cc} T_0 & M_{21}^a & U_0 \\ \hline M_{12}^a & M_{11} & M_{12}^b \\ -V_0 & M_{21}^b & W_0 \end{array} \right],
$$

i.e.,

$$
G_{11} = \left[\begin{array}{c|c} T_0 & M_{21}^a \\ \hline M_{12}^a & M_{11} \end{array} \right]
$$

and so on (compare with (6.23)). The stability of G follows from the assumption that $\det(T_0(s))$ is a Hurwitz polynomial.

Thus by construction, the original transfer function from u to y as given by (6.19), (6.20) is also given by an LFT, namely

$$
\begin{aligned}
G(s; \delta) &= V(s; \delta)T(s; \delta)^{-1}U(s; \delta) + W(s; \delta) \\
&= G_{22}(s) + G_{21}(s)\Delta(I - G_{11}(s)\Delta)^{-1}G_{12}(s) \\
&= \mathcal{F}(\Delta; G(s)).
\end{aligned}
$$

Finally, assume that $G(s; \delta)$ is unstable for some $\Delta_0 \in \Delta$ with $\|\Delta_0\| < \rho^{-1}$. Since the transfer matrices G_{ij} are all stable, this implies that $(I - G_{11}(s)\Delta_0)^{-1}$ has an unstable pole. Thus, there exists an $s_0 \in \overline{\mathbb{C}}_+$ such that $\det(I - G_{11}(s_0)\Delta_0) = 0$. But this implies [97]

$$
\sup_{\omega \in \mathbb{R}} \mu_\Delta(G_{11}(i\omega)) = \sup_{s \in \overline{\mathbb{C}}_+} \mu_\Delta(G_{11}(s)) > \rho,
$$

contradicting the definition of ρ. $\Box$

The given lower bound for the stability radius of $G(s; \delta)$ will, in general, only be useful in case that G_{11} is a *proper* rational function (this guarantees that $\rho < \infty$). Sufficient conditions for properness of G_{11} have been given in special cases of (6.19), (6.20), for instance, descriptor systems [90]; see the subsequent section.

For robust stability issues, it is sufficient to consider the matrix $T(s, \delta)$ alone, rather than the whole coefficient matrix $M(s, \delta)$. For related tasks, such as investigating the influence of certain parameters on the transfer function, an LFT representation of $M(s, \delta)$ may be advantageous.

Note that this approach can be extended without difficulty to the case where $T(s; \delta)$ is a Laurent polynomial with respect to s (this will be the case in the example considered later).

Instead of replacing p_i by δ_i a priori, one can, alternatively, extract first the parameters p_i themselves, and use the fact that the scaling

$$p_i = v_i + m_i \delta_i = \mathcal{F}\left(\delta_i; \begin{bmatrix} 0 & m_i \\ 1 & v_i \end{bmatrix}\right)$$

is itself an LFT. Then common results on the composition of LFTs allow an adaptation of the coefficient matrices obtained for $v = 0$ to an arbitrary nominal point v.

6.4.1 Descriptor Systems

As a special case, let U, V, W be independent of s, and let T be affine linear in s, that is,

$$T(s; \delta) = E(\delta)s - A(\delta).$$

In other words, consider a parametrized linear system in descriptor form

$$E(p)sx = A(p)x + B(p)u \tag{6.25}$$
$$y = C(p)x + D(p)u, \tag{6.26}$$

where with $\mathcal{P} := \mathbb{R}[p]$, the matrices E and A are in $\mathcal{P}^{\nu \times \nu}$, $B \in \mathcal{P}^{\nu \times m}$, $C \in \mathcal{P}^{p \times \nu}$, $D \in \mathcal{P}^{p \times m}$.

As usual, we consider variations around a given nominal parameter point, i.e., we consider

$$E(\delta)sx = A(\delta)x + B(\delta)u \tag{6.27}$$
$$y = C(\delta)x + D(\delta)u \tag{6.28}$$

with nominal parameter point $\delta = 0$. The nominal system is supposed to be stable and of index one. Equivalently, $\det(sE(0) - A(0))$ is assumed to be a Hurwitz polynomial whose degree equals the rank of $E(0)$. In particular, the nominal transfer function

$$G(0) = C(0)(sE(0) - A(0))^{-1}B(0) + D(0)$$

is a proper rational matrix. A comprehensive treatment of singular systems can be found in [9].

Exploiting the structure of T, the procedure described in the previous section can be adapted to the present situation as follows:

1. Define the augmented coefficient matrix

$$M(\delta) = \begin{bmatrix} E(\delta) & A(\delta) & B(\delta) \\ 0 & C(\delta) & D(\delta) \end{bmatrix} \in \mathcal{D}^{g \times q},$$

where $g = \nu + p$ and $q = 2\nu + m$.

2. Write $M(\delta)$ in terms of a linear fractional transformation (LFT)

$$\begin{aligned} M(\delta) &= \mathcal{F}(\Delta; M) \qquad\qquad (6.29) \\ &= M_{22} + M_{21}\Delta(I - M_{11}\Delta)^{-1}M_{12}, \end{aligned}$$

where for some suitable non-negative integers $n_1, \ldots, n_r$ and $n := \sum n_i$,

$$\Delta = \operatorname{diag}(\delta_1 I_{n_1}, \ldots, \delta_r I_{n_r}),$$

$$M = \begin{bmatrix} M_{11} & M_{12} \\ M_{21} & M_{22} \end{bmatrix} \in \mathbb{R}^{(n+g) \times (n+q)}.$$

In this representation, $M_{22} = M(0)$ is the nominal coefficient matrix, i.e., writing $E_0 := E(0)$ and so on, (6.29) takes the form

$$\begin{aligned} M(\delta) = \begin{bmatrix} E_0 & A_0 & B_0 \\ 0 & C_0 & D_0 \end{bmatrix} + \\ \begin{bmatrix} M_{21}^a \\ M_{21}^b \end{bmatrix} \Delta(I - M_{11}\Delta)^{-1} \begin{bmatrix} M_{12}^a & M_{12}^b & M_{12}^c \end{bmatrix}, \end{aligned} \qquad (6.30)$$

where $M_{21} = \begin{bmatrix} M_{21}^a \\ M_{21}^b \end{bmatrix}$ and

$$M_{12} = \begin{bmatrix} M_{12}^a & M_{12}^b & M_{12}^c \end{bmatrix}$$

are partitions that match the corresponding partitions of $M(\delta)$.

3. Define the rational $n \times n$ matrix $G_{11}(s) :=$

$$(M_{12}^b - sM_{12}^a)(sE_0 - A_0)^{-1}M_{21}^a + M_{11} \qquad (6.31)$$

and the set of uncertainties

$$\boldsymbol{\Delta} := \{\Delta = \operatorname{diag}(\delta_1 I_{n_1}, \ldots, \delta_r I_{n_r}), \delta_i \in \mathbb{C}\}.$$

For $s \in i\mathbb{R}$, determine the structured singular value $\mu_{\boldsymbol{\Delta}}(G_{11}(s))$, and let

$$\rho := \sup_{\omega \in \mathbb{R}} \mu_{\boldsymbol{\Delta}}(G_{11}(i\omega)).$$

Corollary 20 *In the situation described above, the system given by (6.27), (6.28) is robustly stable for all parameter variations with $\|\Delta\| < \rho^{-1}$.*

Proof. Plugging the expressions for $E(\delta)$, $A(\delta)$, and $B(\delta)$ from (6.30) into (6.27), one obtains

$$E_0 sx = A_0 x + M_{21}^a \Delta (I - M_{11}\Delta)^{-1} \cdot$$
$$\cdot (M_{12}^b x + M_{12}^c u - M_{12}^a sx) + B_0 u. \tag{6.32}$$

Define

$$z := (I - M_{11}\Delta)^{-1}(M_{12}^b x + M_{12}^c u - M_{12}^a sx),$$
$$w := \Delta z,$$

then (6.32) reads

$$E_0 sx = A_0 x + M_{21}^a w + B_0 u.$$

Similarly, using the LFTs for $C(\delta)$ and $D(\delta)$ from (6.30) in (6.20) and taking into account that by (6.30), $M_{21}^b \Delta (I - M_{11}\Delta)^{-1} M_{12}^a = 0$,

$$\begin{aligned} y &= C_0 x + M_{21}^b \Delta (I - M_{11}\Delta)^{-1} \cdot \\ &\quad \cdot (M_{12}^b x + M_{12}^c u - M_{12}^a sx) + D_0 u \\ &= C_0 x + M_{21}^b w + D_0 u. \end{aligned}$$

Summing up, one obtains the following relations: $w = \Delta z$, and

$$\begin{aligned} E_0 sx &= A_0 x + M_{21}^a w + B_0 u & (6.33) \\ z &= M_{12}^b x + M_{11} w + M_{12}^c u - M_{12}^a sx & (6.34) \\ y &= C_0 x + M_{21}^b w + D_0 u. & (6.35) \end{aligned}$$

The corresponding transfer function from $(w, u)^T$ to $(z, y)^T$ is

$$G = \begin{bmatrix} M_{12}^b - sM_{12}^a \\ C_0 \end{bmatrix} (sE_0 - A_0)^{-1} \begin{bmatrix} M_{21}^a & B_0 \end{bmatrix}$$
$$+ \begin{bmatrix} M_{11} & M_{12}^c \\ M_{21}^b & D_0 \end{bmatrix},$$

i.e., G_{11} as defined in (6.31) corresponds to the transfer from w to z. Thus by construction, the original transfer function from u to y as given by (6.27), (6.28) is

$$\begin{aligned} G(\delta) &= C(\delta)(sE(\delta) - A(\delta))^{-1}B(\delta) + D(\delta) \\ &= G_{22} + G_{21}\Delta(I - G_{11}\Delta)^{-1}G_{12} \\ &= \mathcal{F}(\Delta; G). \end{aligned}$$

The rest follows from Theorem 29. □

As previously remarked, the given lower bound for the stability radius of $G(\delta)$ will, in general, only be useful in case that $G_{11}(s)$ is a *proper* rational function. This can, for instance, be guaranteed if $E(0)$ is non-singular, and only this case is considered in [72]. This requirement is unnecessarily restrictive though, as it will turn out below that we only need the nominal point to be generic in the sense that the rank of $E(\delta)$ as a polynomial matrix does not drop at $\delta = 0$, that is,

$$\operatorname{rank}(E(\delta)) = \operatorname{rank}(E(0)).$$

Note that then without loss of generality, $E(\delta)$ has the following form:

$$E(\delta) = \left[\; E_1(\delta), \;\; 0 \;\right]$$

with a full column rank matrix $E_1(\delta)$ and

$$\operatorname{rank}(E_1(\delta)) = \operatorname{rank}(E_1(0)).$$

Theorem 30 *Let $E(\delta) \in \mathbb{R}[\delta]^{\nu \times \nu}$ be a matrix of the following form:*

$$E(\delta) = \left[\; E_1(\delta), \;\; 0 \;\right],$$

with a full column rank matrix $E_1(\delta) \in \mathbb{R}[\delta]^{\nu \times \nu_1}$ whose rank does not drop at $\delta = 0$, that is, $\operatorname{rank}(E_1(0)) = \operatorname{rank}(E_1(\delta)) = \nu_1$. Then the LFT construction according to Chapter 6 yields

$$\exists X \in \mathbb{R}^{n \times \nu} : \quad M_{12}^a = X E_0,$$

where $E_0 = E(0)$ as usual. Thus one can write $G_{11}(s) =$

$$(M_{12}^b - X A_0)(s E_0 - A_0)^{-1} M_{21}^a + M_{11} - X M_{21}^a$$

from which it is clear that G_{11} is proper rational due to the assumptions on the nominal system.

Proof. By the special structure of $E(\delta)$, we have that the rows of M_{12}^a are either zero or equal to e_i^T for some $1 \le i \le \nu_1$, and each of these e_i^T appears at least once. Thus M_{12}^a can be transformed to reduced echelon form by elementary row transformations, say

$$M_{12}^a = P \begin{bmatrix} I_{\nu_1} & 0 \\ 0 & 0 \end{bmatrix} \tag{6.36}$$

for some non-singular matrix P. On the other hand, by our assumption on E_0, there exists a permutation matrix Π such that

$$\Pi E_0 = \begin{bmatrix} E_{11} & 0 \\ E_{21} & 0 \end{bmatrix}$$

with an invertible $\nu_1 \times \nu_1$-matrix E_{11}. Thus

$$\begin{bmatrix} E_{11}^{-1} & 0 \\ 0 & 0 \end{bmatrix} \Pi E_0 = \begin{bmatrix} I_{\nu_1} & 0 \\ 0 & 0 \end{bmatrix}$$

which, in combination with (6.36), yields the desired result.   □

An analogous argument can be applied to the dual case, where

$$E(\delta) = \begin{bmatrix} E_1(\delta) \\ 0 \end{bmatrix}$$

with a full row rank matrix $E_1(\delta)$ and $\mathrm{rank}\,(E_1(\delta)) = \mathrm{rank}\,(E_1(0))$.

6.4.2 Example

The following matrices provide a linearized description of a common-emitter transistor amplifier [69]. The 6×6 matrix T is given by

$$T = \begin{bmatrix} \frac{p_1}{s} + p_5 + p_6 & -p_5 & 0 \\ -p_5 & \frac{p_2}{s} + p_5 + p_7 & -\frac{p_2}{s} \\ 0 & -\frac{p_2}{s} & \frac{p_2}{s} + p_8 \\ 0 & -p_7 & 0 \\ 0 & 0 & p_8 p_{10} p_{11} \\ 0 & 0 & 0 \end{bmatrix}$$

$$\begin{bmatrix} 0 & 0 & 0 \\ -p_7 & 0 & 0 \\ 0 & 0 & 0 \\ \frac{p_3}{s} + p_7 + p_9 & -\frac{p_3}{s} - p_9 & 0 \\ -\frac{p_3}{s} - p_9 & \frac{p_3}{s} + p_9 + p_{11} + p_{12} & -p_{12} \\ 0 & -p_{12} & \frac{p_4}{s} + p_{12} + p_{13} \end{bmatrix}$$

and the other system matrices are

$$U = [1, 0, 0, 0, 0, 0]^T, \quad V = [0, 0, 0, 0, 0, p_{13}], \quad W = 0.$$

The parameters $p_1, \ldots, p_4$ are reciprocal values of capacitances, p_{10} is a conductance that appears in a voltage-controlled current source, and the remaining parameters are resistances. The nominal values of the 13 parameters are given in the subsequent table.

In spite of the low order of the system and its moderate number of parameters, the denominator of the symbolic transfer function (a polynomial of degree 4 in s) has a support of 225 elements, and it takes more than two pages to print it! Thus, it is hopeless to investigate robust stability at this level.

Table 6.1. Parameter Values.

v_1	v_2	v_3	v_4	v_5
$1.0E4$	$1.8E10$	$2.0E4$	$1.0E4$	$2.9E4$

v_6	v_7	v_8	v_9	v_{10}
$1.0E3$	$5.0E1$	$3.5E3$	$1.0E0$	$6.8E-2$

v_{11}	v_{12}	v_{13}		
$3.4E4$	$7.7E3$	$1.0E3$		

The uncertainty is modeled in a multiplicative way, *i.e.*, $p_i = v_i(1 + \delta_i)$. An LFT representation of the 7×7 coefficient matrix $M(s; \delta)$ is constructed with a 24×24 matrix Δ. The sizes of the repeated blocks corresponding to the uncertainties $\delta = (\delta_1, \ldots, \delta_{13})$ are

$$n = (1, 2, 2, 1, 2, 1, 2, 1, 2, 2, 5, 2, 1).$$

A reduction technique described in Section 6.2 yields an LFT of size 13 which is certainly minimal in the presence of 13 parameters. Thus

$$\Delta = \text{diag}(\delta_1, \ldots, \delta_{13}).$$

The coefficient matrices are

$$M_{12} := \begin{bmatrix}
1 & 0 & 0 & 0 & & 0 & 0 & 0 \\
0 & 1 & -1 & 0 & & 0 & 0 & 0 \\
0 & 0 & 0 & 1 & & -1 & 0 & 0 \\
0 & 0 & 0 & 0 & & 0 & 1 & 0 \\
-1 & 1 & 0 & 0 & & 0 & 0 & 0 \\
-1 & 0 & 0 & 0 & & 0 & 0 & 0 \\
0 & -1 & 0 & 1 & & 0 & 0 & 0 \\
0 & 0 & -1 & 0 & & 0 & 0 & 0 \\
0 & 0 & 0 & -1 & & 1 & 0 & 0 \\
0 & 0 & -1 & 0 & & 0 & 0 & 0 \\
0 & 0 & -1 & 0 & -\frac{1}{v_8 v_{10}} & 0 & 0 \\
0 & 0 & 0 & 0 & & -1 & 1 & 0 \\
0 & 0 & 0 & 0 & & 0 & -1 & 0
\end{bmatrix}$$

$$M_{21} := \begin{bmatrix}
\frac{v_1}{s} & 0 & 0 & 0 & -v_5 & -v_6 \\
0 & \frac{v_2}{s} & 0 & 0 & v_5 & 0 \\
0 & -\frac{v_2}{s} & 0 & 0 & 0 & 0 \\
0 & 0 & \frac{v_3}{s} & 0 & 0 & 0 \\
0 & 0 & -\frac{v_3}{s} & 0 & 0 & 0 \\
0 & 0 & 0 & \frac{v_4}{s} & 0 & 0 \\
0 & 0 & 0 & 0 & 0 & 0
\end{bmatrix}$$

$$\begin{bmatrix} 0 & 0 & 0 & 0 & 0 & 0 & 0 \\ -v_7 & 0 & 0 & 0 & 0 & 0 & 0 \\ 0 & -v_8 & 0 & 0 & 0 & 0 & 0 \\ v_7 & 0 & -v_9 & 0 & 0 & 0 & 0 \\ 0 & -\bar{v} & v_9 & -\bar{v} & -\bar{v} & -v_{12} & 0 \\ 0 & 0 & 0 & 0 & 0 & v_{12} & -v_{13} \\ 0 & 0 & 0 & 0 & 0 & 0 & v_{13} \end{bmatrix}$$

where $\bar{v} := v_8 v_{10} v_{11}$. The matrix M_{11} is a 13×13 matrix with only 3 non-zero entries:

$$M_{11}(10,8) = M_{11}(11,8) = M_{11}(11,10) = 1.$$

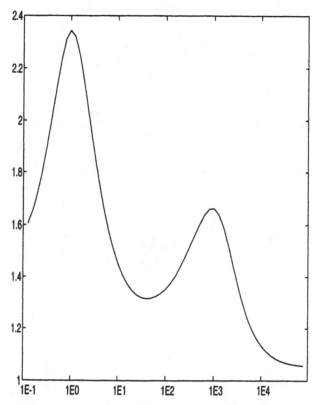

Fig. 6.3. Upper Bound for μ as a Function of Frequency.

It should be noted that the LFT is purely symbolic, *i.e.*, there is no error involved. The figure above shows an upper bound for

$$\mu(\omega) = \mu_{\Delta}(G_{11}(i\omega))$$

as a function of the frequency ω between 10^{-1} and 10^5 rad/ sec, as computed using MATLAB.

Thus $\rho \leq 2.35$, which signifies that 0.42 is a lower bound for the stability radius, *i.e.*, it is guaranteed that the parameters may vary by up to ±42% without making the circuit unstable.

Note that we were able to find an LFT representation whose dimension coincides with the number of uncertain parameters involved, *i.e.*, every δ_i appears in a 1×1 block, or $n = (1, \ldots , 1)$. This is in fact an interesting feature of many commonly used mathematical descriptions of electrical circuits. The subsequent chapter will collect some facts from network theory that are useful for understanding this phenomenon.

7. Electrical Networks

7.1 Graph Theory

This short presentation of basic facts from graph theory is essentially consistent with the nomenclature of Swamy and Thulasiraman [74].

A **graph** $\Gamma = (V, E, \varphi)$ consists of a vertex set V and an edge set E, together with an **incidence map**

$$\varphi : E \to V^2, \quad e \mapsto \varphi(e) = (\varphi_1(e), \varphi_2(e)).$$

If $\varphi(e) = (v_1, v_2)$, one calls v_1 the **initial vertex**, and v_2 the **terminal vertex** of e. Equivalently, edge e is said to be directed from v_1 to v_2. Note that only such **directed** graphs are considered here; moreover, we restrict to **finite** graphs, that is, both V and E are supposed to be finite sets. A **self-loop** is an edge e with $\varphi_1(e) = \varphi_2(e)$. It is useful to define the following sets of edges associated with a vertex $v \in V$: Let

$$E(v, -) = \varphi_1^{-1}(v) = \{e \in E, \varphi_1(e) = v\} \quad \text{and}$$
$$E(-, v) = \varphi_2^{-1}(v) = \{e \in E, \varphi_2(e) = v\}$$

denote the sets of all edges with initial or terminal vertex v, respectively. The cardinalities of these sets,

$$d_{out}(v) = |E(v, -)| \quad \text{and} \quad d_{in}(v) = |E(-, v)|$$

are called the **out-degree** and **in-degree** of $v \in V$, respectively. Finally, $d(v) = d_{in}(v) + d_{out}(v)$ is the total number of edges incident on v, and it is called the **degree** of v.

For each edge e, define an additional edge $-e$ that is directed from the terminal to the initial vertex of e, that is, $\varphi(-e) := (\varphi_2(e), \varphi_1(e))$ for $e \in E$. Define $\tilde{E} = E \dot{\cup} (-E)$. A **path** is a sequence

$$\underline{w} = (w_1, \ldots, w_r) \in \tilde{E}^r$$

of length $r \geq 0$, where for $1 \leq i \leq r - 1$,

$$\varphi_2(w_i) = \varphi_1(w_{i+1}) =: v_{k_i}.$$

Vertex $v_{k_0} := \varphi_1(w_1)$ is called the initial, vertex $v_{k_r} := \varphi_2(w_r)$ is called the terminal vertex of $\underline{w}$. Equivalently, $\underline{w}$ is said to be a path from v_{k_0} to v_{k_r}. An **elementary** path is one in which $v_{k_1}, \dots, v_{k_r}$ are distinct. A **circuit** is an elementary path with $v_{k_0} = v_{k_r}$. Two vertices v, v' are said to be **connected** iff there exists a path from v to v'. The graph itself is called connected if any two vertices are connected.

A **spanning subgraph** of $\Gamma = (V, E, \varphi)$ is a graph $\Gamma' = (V, E', \varphi|_{E'})$, where $E' \subseteq E$. Usually, it is the edge set E' itself that is referred to as spanning subgraph. A **proper** spanning subgraph is one in which $E' \neq E$. A **tree** is a minimally connected graph, *i.e.*, it is connected without having any proper spanning subgraphs that are connected. In a tree with vertex set V and edge set E, we must have that

$$|E| = |V| - 1.$$

A spanning subgraph E' of Γ that is a tree is called a **spanning tree** of Γ. Its complement $E \setminus E'$ is called **co-tree**. Every connected graph possesses a spanning tree.

7.2 Weight Functions

Let R be an arbitrary ring with unity. Vertex or edge **weight functions** are mappings $V \to R$ or $E \to R$, respectively, *i.e.*, they assign an element of R to each vertex or edge of the graph. The main cases that should be thought of are $R = \mathbb{Z}$ (important for topological considerations), and R being a field, e.g., $\mathbb{R}$ or $\mathbb{C}$. With electrical networks, R usually denotes the field of rational or, more generally, of meromorphic functions defined on the open right half of the complex plane. The **boundary operator** [68]

$$\partial : R^E \to R^V, \quad i \mapsto \partial(i)$$

assigns a vertex weight function to each edge weight function $i : E \to R$ in the following fashion:

$$\partial(i)(v) = \sum_{e \in E(v,-)} i(e) - \sum_{e \in E(-,v)} i(e).$$

In the case $R = \mathbb{R}$, one may think of i as a flow, the value $i(e)$ being the amount of a material that is transported along the edge e, negative values corresponding to transport in reverse direction, *i.e.*, along $-e$. Then $\partial(i)(v)$ is the net outflow of i at vertex v. The dual mapping

$$\partial^* : R^V \cong \mathrm{Hom}_R(R^V, R) \to \mathrm{Hom}_R(R^E, R) \cong R^E, \quad \phi \mapsto \partial^*(\phi) = \phi \circ \partial$$

is given by $\partial^*(\phi)(e) = \phi(\varphi_1(e)) - \phi(\varphi_2(e))$, that is, $\partial^*(\phi)$ assigns to each edge the weight difference between its initial and terminal vertex. The map ∂^* is also the adjoint of ∂ with respect to the standard bilinear functions defined by

$$\langle \cdot, \cdot \rangle_E : R^E \times R^E \to R, \qquad (i,j) \mapsto \langle i,j \rangle_E := \sum_{e \in E} i(e)j(e)$$

$$\langle \cdot, \cdot \rangle_V : R^V \times R^V \to R \qquad (\phi, \psi) \mapsto \langle \phi, \psi \rangle_V := \sum_{v \in V} \phi(v)\psi(v).$$

That is, we have

$$
\begin{aligned}
\langle \partial(i), \phi \rangle_V &= \sum_{v \in V} \left(\sum_{e \in \varphi_1^{-1}(v)} i(e) - \sum_{e \in \varphi_2^{-1}(v)} i(e) \right) \phi(v) \\
&= \sum_{e \in E} i(e)\phi(\varphi_1(e)) - \sum_{e \in E} i(e)\phi(\varphi_2(e)) = \langle i, \partial^*(\phi) \rangle_E
\end{aligned}
$$

for all $i \in R^E$ and $\phi \in R^V$.

7.3 Kirchhoff's Laws

An electrical network is given by a connected graph and two edge weight functions i and u, interpreted as currents and voltages along the edges of the network. As we are going to consider the dynamics of the network in the frequency domain, the values of these edge weight functions are Laplace transforms of the time domain signals

$$i, u : [0, \infty) \to \mathbb{K},$$

where $\mathbb{K} = \mathbb{R}$ or $\mathbb{K} = \mathbb{C}$. A typical representative of the class of time domain signals considered here is the sinusoid

$$a \cos(\omega t) + b \sin(\omega t)$$

where $a, b, \omega \in \mathbb{R}$, with Laplace transform $\frac{as+b\omega}{s^2+\omega^2}$. In particular, the frequency domain signals are analytic in $\mathbb{C}_+ = \{s \in \mathbb{C}, \text{Re}(s) > 0\}$. Thus, one can think of the co-domain R of the edge weight functions i, u as the ring of analytic functions on $\mathbb{C}_+$. Later on, it will be more convenient to work with fields. It is well-known that the ring of analytic functions on $\mathbb{C}_+$ is an integral domain. Therefore it can be embedded into its quotient field, the field of meromorphic functions on $\mathbb{C}_+$. If we restrict to sinusoidal time domain signals, it suffices to work with the field of rational functions. However, note that the considerations below apply to arbitrary rings R.

Kirchhoff's current law (KCL) says that the net current outflow vanishes at any vertex of the graph, that is, $\partial(i) \equiv 0$, or,

$$i \in \ker(\partial).$$

According to **Kirchhoff's voltage law** (KVL), voltages are potential differences, *i.e.*, u can be derived from an electrical potential $\phi \in R^V$ via $u = \partial^*(\phi)$. In other words,

$$u \in \operatorname{im}(\partial^*).$$

An immediate consequence of KCL and KVL is known as **Tellegen's theorem** [10, 74]: It states orthogonality of current and voltage with respect to $\langle \cdot, \cdot \rangle_E$. This is due to

$$\langle i, u \rangle_E = \langle i, \partial^*(\phi) \rangle_E = \langle \partial(i), \phi \rangle_V = \langle 0, \phi \rangle_V = 0.$$

Let Γ be a connected graph. Then the following sequence is exact:

$$0 \to \ker(\partial) \hookrightarrow R^E \xrightarrow{\partial} R^V \xrightarrow{\pi} R \to 0, \qquad (7.1)$$

where $\pi : R^V \to R$ is defined by

$$\pi(\phi) = \sum_{v \in V} \phi(v).$$

One can show that $\ker(\partial)$ is again a finite free R-module (this is clear when R is a principal ideal domain or even a field, but holds over arbitrary rings as well). In particular, we have

$$\operatorname{rank}(\partial) = \operatorname{rank}(R^V) - \operatorname{rank}(\pi) = |V| - 1$$

and thus

$$\operatorname{rank} \ker(\partial) = \operatorname{rank}(R^E) - \operatorname{rank}(\partial) = |E| - |V| + 1.$$

In the following, let $V = \{v_1, \dots, v_n\}$ and $E = \{e_1, \dots, e_m\}$, and identify $R^E = R^m$, and $R^V = R^n$. The natural matrix representation of ∂ is given by $A = (a_{ij})$ with

$$a_{ij} = \begin{cases} 1 & \text{if } \varphi_1(e_j) = v_i \text{ and } \varphi_2(e_j) \neq v_i \\ -1 & \text{if } \varphi_2(e_j) = v_i \text{ and } \varphi_1(e_j) \neq v_i \\ 0 & \text{otherwise.} \end{cases}$$

This matrix is known as (all-vertex) **incidence matrix** of Γ. Due to the fact that the sum of the rows of A yields zero, *i.e.*,

$$\begin{bmatrix} 1 & \cdots & 1 \end{bmatrix} A = 0,$$

there is no loss of information in deleting an arbitrarily chosen row of A. In electrical engineering, it is usually the resulting reduced full row rank matrix that is called incidence matrix, but we will not stick to this nomenclature here.

A basis of ker(∂) can be found by taking a spanning tree of Γ. Then for every edge in the co-tree, say e with $\varphi(e) = (v_1, v_2)$, there exists one and only one elementary path from v_2 to v_1 in the tree. Together with e, this constitutes a circuit. The circuits constructed like this are called **fundamental circuits** with respect to the chosen spanning tree. An edge weight function $b \in R^m$ is associated with each fundamental circuit $\underline{w} = (w_1, \ldots, w_r)$ via

$$b_k = \begin{cases} 1 & \text{if } \exists j : w_j = e_k \in E \\ -1 & \text{if } \exists j : w_j = -e_k \in -E \\ 0 & \text{otherwise.} \end{cases}$$

As the spanning tree contains $n-1$ edges, there are $l := m-(n-1)$ fundamental circuits. It can be shown that the l edge weight functions defined above span ker(∂). As rank ker(∂) $= l$, they are indeed a basis of ker(∂). Arranging these basis vectors column-wise in a matrix B^T, one may identify $R^l \cong \ker \partial$, $x \leftrightarrow B^T x$. The natural matrix representation of π is $\begin{bmatrix} 1 & \ldots & 1 \end{bmatrix} \in R^{1 \times n}$. Thus the exact sequence (7.1) can be replaced by its equivalent counterpart

$$0 \to R^l \xrightarrow{B^T} R^m \xrightarrow{A} R^n \xrightarrow{[1\ldots1]} R \to 0.$$

The incidence matrix $A \in R^{n \times m}$, the circuit matrix $B \in R^{l \times m}$, and the exactness of $AB^T = 0$, that is, the fact that

$$\operatorname{rank}(A) + \operatorname{rank}(B) = m,$$

give rise to the following equivalent formulations of KVL:

1. $Ai = 0$;
2. There exists $j \in R^l$ (**circuit current**) such that $i = B^T j$.

Similarly, KVL can be expressed in two equivalent ways:

1. There exists $\phi \in R^n$ (**node potential**) such that $u = A^T \phi$;
2. $Bu = 0$.

7.4 Linear Networks

In an electrical network, the current i and the voltage u are not only subject to Kirchhoff's laws, but also to additional relations, which, in the linear case, can be put in the form

$$Pi + Qu = Ri_0 + Su_0 \tag{7.2}$$

for some given coefficient matrices P, Q, R, S and given i_0, u_0. Typical relations are $Ri_k - u_k = 0$ (resistor), $Lsi_k - u_k = 0$ (inductor), $i_k - Csu_k = 0$

(capacitor), $u_k = u_{0j}$ or $i_k = i_{0j}$ (free voltage or current source). These network elements all lead to diagonal matrices P and Q, as they relate the current and voltage along one and the same edge of the graph. There are, however, also networks elements that destroy the diagonal structure of the coefficient matrices, for instance, voltage-controlled current sources (VCCS): $i_k - gu_j = 0$ with $k \neq j$, or current-controlled current sources (CCCS): $i_k - \beta i_j = 0$, $k \neq j$. These controlled sources arise in the linearization of non-linear network elements such as transistors. Inductors and capacitors represent memory-possessing, dynamic i-u-relations; they are also called **reactive** elements. Networks without reactive elements are said to be **resistive**, although they are not necessarily composed solely of resistors.

Recall that i, u belong to the field $\mathcal{M}$ of meromorphic functions on $\mathbb{C}_+$. This applies also to the given right-hand-side signals, i_0, u_0. The base field of the coefficient matrices can be $\mathbb{K} = \mathbb{R}, \mathbb{C}$ (resistive network with numeric parameter values), $\mathbb{K}(s)$ (general linear network with numeric parameter values), or, in the most general situation, $\mathbb{K}(s; p_1, \dots, p_r)$ (symbolic parameter values), depending on the particular problem. The indeterminates $p_1, \dots, p_r$ correspond to (a subset of) the appearing network parameters such as resistances, admittances, capacitances, inductances, and so on. All of these base fields operate in the natural way on $\mathcal{M}$ or $\mathcal{M}(p_1, \dots, p_r)$, respectively.

Summing up, the circuit equations can be written in the following form known as the **sparse tableau**:

$$\begin{bmatrix} A & 0 \\ 0 & B \\ P & Q \end{bmatrix} \begin{bmatrix} i \\ u \end{bmatrix} = \begin{bmatrix} 0 & 0 \\ 0 & 0 \\ R & S \end{bmatrix} \begin{bmatrix} i_0 \\ u_0 \end{bmatrix}. \tag{7.3}$$

Introducing the circuit current j and the node potential ϕ according to Kirchhoff's laws, we obtain the following compressed version of (7.3) with $m + 1$ instead of $2m$ unknowns:

$$\begin{bmatrix} PB^T & QA^T \end{bmatrix} \begin{bmatrix} j \\ \phi \end{bmatrix} = \begin{bmatrix} R & S \end{bmatrix} \begin{bmatrix} i_0 \\ u_0 \end{bmatrix}. \tag{7.4}$$

It will be assumed throughout the following discussion that these equations are sufficient for determining i and u uniquely. The coefficient matrices P and Q are supposed to be square. Recalling that a row of A can be eliminated without loss of information, the assumption boils down to

$$\det \begin{bmatrix} A' & 0 \\ 0 & B \\ P & Q \end{bmatrix} \neq 0, \quad \text{or} \quad \det \begin{bmatrix} PB^T & QA'^T \end{bmatrix} \neq 0,$$

respectively, where A' denotes A after deletion of an arbitrarily chosen row.

For a special class of circuits, namely RCL–networks with (free or voltage-controlled) current sources, the **nodal analysis** method makes it possible to reduce the system even further. For networks of this type, we have that the matrix P is invertible. (In fact, one can even assume $P = I_m$ without loss of generality.) Then the equations $Pi + Qu = c$, where $c := Ri_0 + Su_0$, can be put in the form

$$i + Yu = d,$$

where $Y := P^{-1}Q$ denotes the **admittance matrix** and $d := P^{-1}c$. Multiply these equations by A from the left to obtain, using $Ai = 0$ and $u = A^T\phi$,

$$AYA^T\phi = Ad. \tag{7.5}$$

Note that (7.5) contains the same information as the sparse tableau: Solvability of (7.5) is guaranteed by the fact that

$$\text{rank}\,(AYA^T) = \text{rank}\,(A), \tag{7.6}$$

and thus $\text{im}(AYA^T) = \text{im}(A)$. From the nodal potential ϕ, voltage and current can be reconstructed using $u = A^T\phi$ and $i = d - Yu$. To see that (7.6) is true, consider

$$A'P^{-1}\begin{bmatrix} PB^T & QA'^T \end{bmatrix} = \begin{bmatrix} 0 & A'YA'^T \end{bmatrix}.$$

Hence $\text{rank}\,(A'YA'^T) = \text{rank}\,(A') = n - 1$. Thus AYA^T contains a nonsingular sub-matrix of dimension $n - 1$. On the other hand, as all the rows and columns of AYA^T add up to zero, its rank can be at most $n - 1$, hence $\text{rank}\,(AYA^T) = n - 1$ as desired. Equation (7.5) is under-determined in the sense that ϕ is unique up to an additive constant. Usually, a reference node potential is set to be zero, thus replacing (7.4) or (7.5) by the regular systems

$$\begin{bmatrix} PB^T & QA'^T \end{bmatrix}\begin{bmatrix} j \\ \phi' \end{bmatrix} = c, \quad \text{or} \quad A'YA'^T\phi' = A'd, \tag{7.7}$$

respectively. For theoretical considerations however, the full set of equations (7.5) is preferable due to the structure of its coefficient matrix $Y_{\text{IAM}} = AYA^T$, which is called **indefinite admittance matrix** (IAM). The entries of Y appear in Y_{IAM} in characteristic patterns: Decompose $Y = (y_{ij})$ as

$$Y = \sum_{i,j=1}^{m} y_{ij}e_ie_j^T,$$

where e_i denotes the i-th natural basis vector. Then

$$AYA^T = \sum_{i,j} y_{ij}Ae_ie_j^TA^T = \sum_{i,j} y_{ij}A_{-i}(A_{-j})^T$$

where A_{-i} denotes the i-th column of A, which is of the form

$$A_{-i} = e_{\mu_i} - e_{\nu_i}$$

if the initial and terminal vertex of edge e_i are v_{μ_i} and v_{ν_i}, respectively. Assume without loss of generality $\mu_i < \nu_i$ (if $\mu_i = \nu_i$, edge e_i is a self-loop, and the i-th column of A is zero) and $\mu_j < \nu_j$, then $y_{ij} A_{-i} A_{-j}^T$ takes the following form:

$$
\begin{array}{cc}
\overset{\displaystyle \mu_j}{\mid} & \overset{\displaystyle \nu_j}{\mid}
\end{array}
$$

$$
\begin{array}{c}
\mu_i \;- \\[2em]
\nu_i \;-
\end{array}
\left[
\begin{array}{cc}
y_{ij} & -y_{ij} \\[2em]
-y_{ij} & y_{ij}
\end{array}
\right]
$$

and Y_{IAM} is a superposition of matrices with this structure.

Furthermore, Y_{IAM} is an **equi-cofactor** matrix, *i.e.*, all its cofactors are equal. Let $\mathrm{cof}_{ij}(Y_{IAM})$ denote the (i,j)-cofactor of Y_{IAM}, that is $(-1)^{i+j}$ times the sub-determinant of Y_{IAM} after deletion of the i-th row and j-th column. To see that the cofactors are all equal, consider the system of equations $Y_{IAM} x = b$ with given right hand side $b = e_\mu - e_\nu$. As the column sum of b is zero, *i.e.*, $b \in \ker[1, \ldots , 1] = \mathrm{im}(A) = \mathrm{im}(Y_{IAM})$, this system of equations is solvable. Hence Cramer's rule

$$\det(Y_{IAM}) x_j = \sum_{i=1}^{n} b_i \mathrm{cof}_{ij}(Y_{IAM})$$

is applicable, and it implies

$$0 = \mathrm{cof}_{\mu j}(Y_{IAM}) - \mathrm{cof}_{\nu j}(Y_{IAM}).$$

This holds for all indices μ, ν, j, thus showing that the cofactors are constant along the columns of Y_{IAM}. A similar argument can be applied to the rows. In particular, for any choice of the reference node potential, the cofactor of Y_{IAM} is

$$\mathrm{cof}(A Y A^T) = \det(A' Y A'^T) \neq 0.$$

If Y is a diagonal matrix, say $Y = \mathrm{diag}(Y_1, \ldots , Y_m)$ (this is true for RCL-networks with free current sources), the Cauchy-Binet theorem [27] implies that $\det(A' Y A'^T) =$

$$\sum_{i_1 < \ldots < i_{n-1} \leq m} \det(A'(i_1, \ldots , i_{n-1})) Y_{i_1} \cdots Y_{i_{n-1}} \det(A'(i_1, \ldots , i_{n-1})),$$

where $A'(i_1, \ldots , i_{n-1})$ denotes the sub-matrix of A' which consists of the columns with these indices. Now these sub-determinants can only take the values $-1, 0, 1$. The case ± 1 corresponds to a spanning tree of the graph. Thus

$$\det(A'YA'^T) = \sum_{i_1,\dots,i_{n-1}\text{tree}} Y_{i_1} \cdots Y_{i_{n-1}},$$

where the summation runs over all spanning trees $\{e_{i_1},\dots,e_{i_{n-1}}\}$ of the graph Γ. The number of spanning trees of Γ is given by $\det(A'A'^T)$.

These nice features of the indefinite admittance matrix Y_{IAM} are partially lost in the case where P in (7.2) is not invertible. Still one can perform the so-called **modified nodal analysis (MNA)**, a technique that covers a broad class of electrical networks. The MNA method is implemented in several program packages for circuit analysis, such as SPICE or ANALOG INSYDES. It is based on a splitting of the circuit equations (7.2) into a part that can be solved for i (as with the standard nodal analysis) and a part that can be solved for u. Indeed, suppose that (7.2) takes the form

$$\begin{bmatrix} I & P_{12} \\ 0 & P_{22} \end{bmatrix} \begin{bmatrix} i_1 \\ i_2 \end{bmatrix} + \begin{bmatrix} Q_{11} & 0 \\ Q_{21} & I \end{bmatrix} \begin{bmatrix} u_1 \\ u_2 \end{bmatrix} = \begin{bmatrix} c_1 \\ c_2 \end{bmatrix}.$$

With the corresponding partition, we obtain from Kirchhoff's laws,

$$A'_1 i_1 + A'_2 i_2 = 0 \quad \text{and} \quad \begin{bmatrix} u_1 \\ u_2 \end{bmatrix} = \begin{bmatrix} A_1'^T \\ A_2'^T \end{bmatrix} \phi.$$

This yields the MNA equations

$$\begin{bmatrix} A'_1 Q_{11} A_1'^T & A'_1 P_{12} - A'_2 \\ Q_{21} A_1'^T + A_2'^T & P_{22} \end{bmatrix} \begin{bmatrix} \phi \\ i_2 \end{bmatrix} = \begin{bmatrix} A'_1 c_1 \\ c_2 \end{bmatrix}. \tag{7.8}$$

Note that the system matrix takes the form

$$\begin{bmatrix} A'_1 & 0 \\ 0 & I \end{bmatrix} \begin{bmatrix} Q_{11} & P_{12} \\ Q_{21} & P_{22} \end{bmatrix} \begin{bmatrix} A'_1 & 0 \\ 0 & I \end{bmatrix}^T + \begin{bmatrix} 0 & -A'_2 \\ A_2'^T & 0 \end{bmatrix}$$

or $LXL^T + K$ with matrices L, K whose entries are in $\{-1, 0, 1\}$, and a mixed admittance/impedance matrix

$$X = \begin{bmatrix} Q_{11} & P_{12} \\ Q_{21} & P_{22} \end{bmatrix}$$

(in terms of physical units, Q_{11} is an admittance, P_{22} is an impedance, and P_{12}, Q_{21} are dimensionless).

Reduction to first order: Consider the regular representations of the circuit equations given in (7.7) or (7.8). They can be put in the general form

$$T(s; p)\xi(s) = U(p)\nu(s), \tag{7.9}$$

where $p = (p_1, \ldots, p_r)$ is the vector of network parameters, $\xi(s)$ denotes the generalized state (e.g., $\xi \equiv \phi'$), and ν is the vector of inputs (free sources) that enters the right hand sides of (7.7), (7.8) via some transformation U that is assumed to be independent of s. The matrix T is invertible over $\mathbb{K}(s; p)$. Due to the structure of the admissible current-voltage-relations, T splits into three parts:

$$T(s; p) = T_1(p)s + T_0(p) + T_{-1}(p)\frac{1}{s}.$$

Equation (7.9) can be reduced to first order by introducing the enlarged state vector

$$x(s) = \left[\begin{array}{c} \xi(s)/s \\ \xi(s) \end{array} \right]$$

leading to the parametrized descriptor system [9, 72, 92]

$$E(p)sx(s) = A(p)x(s) + B(p)\nu(s) \tag{7.10}$$

where

$$E(p) = \left[\begin{array}{cc} I & 0 \\ 0 & T_1(p) \end{array} \right], \; A(p) = \left[\begin{array}{cc} 0 & I \\ -T_{-1}(p) & -T_0(p) \end{array} \right], \; B(p) = \left[\begin{array}{c} 0 \\ U(p) \end{array} \right].$$

7.5 LFT Representations

Consider an electrical circuit whose system matrix – evaluated at a certain frequency $s = i\omega$, $\omega \in \mathbb{R}$ – can be set up by a modified nodal analysis (MNA):

$$P = LXL^T + K, \tag{7.11}$$

where L and K take their values in $\{-1, 0, 1\}$. Admitting resistors, inductors, capacitors, and all types of free and controlled sources, there is no loss of generality in assuming that X depends on the network parameters $p_1, \ldots, p_r$ as follows:

$$X = \sum_{i=1}^{r} p_i e_{\rho(i)} e_{\gamma(i)}^T. \tag{7.12}$$

Here e_j denotes the j-th natural basis vector, and for each i, $\rho(i)$ and $\gamma(i)$ denote the row and column index of p_i in X, respectively.

Now let each parameter $p_i = v_i + m_i \delta_i$ be subject to uncertainty. Let

$$V = \sum_{i=1}^{r} v_i e_{\rho(i)} e_{\gamma(i)}^T, \quad \Delta = \text{diag}(\delta_1, \ldots, \delta_r), \quad M = \text{diag}(m_1, \ldots, m_r).$$

Define

$$N_L = \begin{bmatrix} e_{\rho(1)} & \cdots & e_{\rho(r)} \end{bmatrix} \quad \text{and} \quad N_R = \begin{bmatrix} e_{\gamma(1)}^T \\ \vdots \\ e_{\gamma(r)}^T \end{bmatrix}.$$

Then

$$X = V + N_L M \Delta N_R$$

and thus

$$P = LXL^T + K = LVL^T + LN_L M \Delta N_R L^T + K.$$

Thus we have written P in LFT form, $P = \mathcal{F}(\Delta; \tilde{A}, \tilde{B}, \tilde{C}, \tilde{D})$ with

$$\tilde{A} = 0, \quad \tilde{B} = N_R L^T, \quad \tilde{C} = LN_L M, \quad \tilde{D} = LVL^T + K.$$

A well-known matrix inversion formula (the Bartlett-Sherman-Morrison-Woodbury formula, see [31] for a historic account) yields

$$P^{-1} = \tilde{D}^{-1} - \tilde{D}^{-1}\tilde{C}\Delta(I - (\tilde{A} - \tilde{B}\tilde{D}^{-1}\tilde{C})\Delta)^{-1}\tilde{B}\tilde{D}^{-1}.$$

Define

$$\begin{aligned} A &:= \tilde{A} - \tilde{B}\tilde{D}^{-1}\tilde{C} \\ B &:= \tilde{B}\tilde{D}^{-1} \\ C &:= -\tilde{D}^{-1}\tilde{C} \\ D &:= \tilde{D}^{-1}, \end{aligned}$$

then $P^{-1} = \mathcal{F}(\Delta; A, B, C, D)$, that is, we have found a representation of P^{-1} that involves only the inversion of the nominal matrix $P|_{\delta=0} = \tilde{D}$. Define $A_1 := N_R L^T (LVL^T + K)^{-1} LN_L$, then

$$A = -A_1 M = -N_R L^T (LVL^T + K)^{-1} LN_L M, \tag{7.13}$$

and

$$B = N_R L^T (LVL^T + K)^{-1} \quad \text{and} \quad C = -(LVL^T + K)^{-1} LN_L M. \tag{7.14}$$

In particular, both P and its inverse can be represented through LFTs in which the block partition of Δ is $n = (1, \dots, 1)$. This is due to the structure of X assumed in (7.12).

Next, we take a closer look at the balancing notion of Section 6.3 applied to this special case: As A is a multiple of M according to (7.13), it can be made $(1, \dots, 1)$-stable by proper choice of M: Compute $\mu \approx \mu(A_1)$ (compare (6.14)) and a diagonal matrix D such that $\|DA_1 D^{-1}\|_2 = \mu$ (both can be done using the MATLAB μ-Analysis and Synthesis Toolbox). Then for any M with $\|M\|_2 < \mu^{-1}$, we have

$$\|DAD^{-1}\|_2 = \|DA_1MD^{-1}\|_2 = \|DA_1D^{-1}M\|_2 \leq \|DA_1D^{-1}\|_2\,\|M\|_2 < 1.$$

Then

$$AD^{-2}A^* - D^{-2} < 0 \quad \text{and} \quad A^*D^2A - D^2 < 0.$$

As in (6.15), let

$$\lambda_B > \rho(BB^*(D^{-2} - AD^{-2}A^*)^{-1}) \quad \text{and} \quad \lambda_C > \rho(C^*C(D^2 - A^*D^2A)^{-1}).$$

Then the construction following Definition 42 yields $\Sigma = \sqrt{\lambda_B\lambda_C}I$. Of course, this Gramian is not very useful from the point of view of balancing. It just reflects the fact that all the 1×1 blocks, each corresponding to one parameter, are treated equally. But recall that Σ is not unique! In fact, we can divide the parameters into two disjoint subsets of parameters "to-be-kept" and parameters "to-be-eliminated," say without loss of generality, $\{p_1, \ldots, p_k\}$ and $\{p_{k+1}, \ldots, p_r\}$, respectively. Then our objective will be to find a solution to $APA^* - P + BB^* < 0$ of the form

$$P = \mathrm{diag}(\pi_1, \ldots, \pi_k, \pi_{k+1}, \ldots, \pi_r)$$

such that $\pi_{k+1} + \ldots + \pi_r \to \min$. Similarly, we will look for a solution Q to $A^*QA - Q + C^*C < 0$ with block structure

$$Q = \mathrm{diag}(q_1, \ldots, q_k, q_{k+1}, \ldots, q_r)$$

such that $q_{k+1} + \ldots + q_r \to \min$. Problems like that are solved by the MATLAB LMI Control Toolbox. If additionally, the first k diagonal entries of P and Q are required to be equal, the balanced truncation is nothing but a truncation of the original system, with $p_{k+1}, \ldots, p_r$ set to their respective nominal values.

To get an initial guess on which parameters to eliminate, it is useful to consider the sensitivity of the nominal model with respect to each parameter: Let $P = LXL^T + K$ and $Q = P^{-1}$. Define $Q_{\mathrm{nom}} := Q|_{p=v} = (LVL^T + K)^{-1}$. We define the **sensitivity** of Q with respect to the i-th parameter by

$$S_i := \left.\frac{\partial Q}{\partial p_i}\right|_{p=v} = \left.-Q\frac{\partial P}{\partial p_i}Q\right|_{p=v} = \left.-Q_{\mathrm{nom}}L\frac{\partial X}{\partial p_i}\right|_{p=v}L^TQ_{\mathrm{nom}}.$$

Using $X = \sum_{i=1}^{r} p_i e_{\rho(i)} e_{\gamma(i)}^T$,

$$S_i = -Q_{\mathrm{nom}}Le_{\rho(i)}e_{\gamma(i)}^T L^TQ_{\mathrm{nom}}. \tag{7.15}$$

From the representation $Q = D + C\Delta(I - A\Delta)^{-1}B$ and $p_i = v_i + m_i\delta_i$, we obtain

$$S_i = C\frac{e_i e_i^T}{m_i}B \tag{7.16}$$

as an alternative expression for the sensitivity of Q with respect to p_i. Using the expressions for B and C derived in (7.14), it is easy to see that (7.15) and (7.16) coincide, noting that $Q_{\text{nom}} = D = (LVL^T + K)^{-1}$.

Example 7.5.1. The network represented in the figure below consists of 4 nodes and 7 edges; a reference node potential has been chosen as indicated. The equations depend on 5 parameters, denoted by $G_1, \ldots, G_4$ and g_m, and they admit a standard nodal analysis.

We have

$$Y = \begin{bmatrix} G_1 & 0 & 0 & 0 & 0 & 0 & 0 \\ 0 & G_2 & 0 & 0 & 0 & 0 & 0 \\ 0 & 0 & G_3 & 0 & 0 & 0 & 0 \\ 0 & 0 & 0 & G_4 & 0 & 0 & 0 \\ 0 & 0 & 0 & 0 & 0 & 0 & 0 \\ 0 & 0 & 0 & 0 & 0 & 0 & 0 \\ 0 & 0 & 0 & 0 & 0 & g_m & 0 \end{bmatrix} = G_1 e_1 e_1^T + \ldots + G_4 e_4 e_4^T + g_m e_7 e_6^T$$

$$N_L = \begin{bmatrix} 1 & 0 & 0 & 0 & 0 \\ 0 & 1 & 0 & 0 & 0 \\ 0 & 0 & 1 & 0 & 0 \\ 0 & 0 & 0 & 1 & 0 \\ 0 & 0 & 0 & 0 & 0 \\ 0 & 0 & 0 & 0 & 0 \\ 0 & 0 & 0 & 0 & 1 \end{bmatrix} \qquad N_R = \begin{bmatrix} 1 & 0 & 0 & 0 & 0 & 0 & 0 \\ 0 & 1 & 0 & 0 & 0 & 0 & 0 \\ 0 & 0 & 1 & 0 & 0 & 0 & 0 \\ 0 & 0 & 0 & 1 & 0 & 0 & 0 \\ 0 & 0 & 0 & 0 & 0 & 1 & 0 \end{bmatrix}$$

The reduced incidence matrix is

$$J = \begin{bmatrix} 1 & 0 & 0 & 1 & -1 & 1 & 0 \\ 0 & 0 & 1 & -1 & 0 & -1 & 1 \\ 0 & 1 & -1 & 0 & 0 & 0 & -1 \end{bmatrix}$$

This yields the admittance matrix

$$P = JYJ^T = \begin{bmatrix} G_1 + G_4 & -G_4 & 0 \\ g_m - G_4 & G_3 + G_4 - g_m & -G_3 \\ -g_m & g_m - G_3 & G_2 + G_3 \end{bmatrix}$$

As an academic example, let all nominal values be equal to one, *i.e.*,

$$V = \begin{bmatrix} 1 & 0 & 0 & 0 & 0 & 0 & 0 \\ 0 & 1 & 0 & 0 & 0 & 0 & 0 \\ 0 & 0 & 1 & 0 & 0 & 0 & 0 \\ 0 & 0 & 0 & 1 & 0 & 0 & 0 \\ 0 & 0 & 0 & 0 & 0 & 0 & 0 \\ 0 & 0 & 0 & 0 & 0 & 0 & 0 \\ 0 & 0 & 0 & 0 & 0 & 1 & 0 \end{bmatrix}$$

As $Y = V + N_L M \Delta N_R$, we have

$$P = JYJ^T = JVJ^T + JN_L M \Delta N_R J^T$$

and hence $\tilde{A} = 0$, $\tilde{B} = N_R J^T$, $\tilde{C} = JN_L M$, $\tilde{D} = JVJ^T$.

$$A_1 = N_R J^T (JVJ^T)^{-1} JN_L = \frac{1}{3} \begin{bmatrix} 2 & 1 & 1 & 0 & 1 \\ 1 & 2 & -1 & 0 & -1 \\ 0 & 0 & 3 & -3 & 3 \\ 1 & -1 & -1 & 3 & -1 \\ 1 & -1 & -1 & 3 & -1 \end{bmatrix}$$

We have $\rho(A_1) = 1$ and $\|A_1\|_2 = \frac{7}{3}$. Using MATLAB, $\mu(A_1) \approx \mu = 2.1547$ and $D = \text{diag}(1, 1.3149, 0.7605, 1, 1)$ is such that $\|DA_1 D^{-1}\|_2 = \mu$. In fact, an inspired guess yields the exact expressions $\mu(A_1) = \frac{2}{3}\sqrt{3} + 1$ and $D = \text{diag}(1, 3^{\frac{1}{4}}, 3^{-\frac{1}{4}}, 1, 1)$. Put $M = \frac{1}{10}I$, this guarantees that $A = -A_1 M$ is $(1, \dots, 1)$-stable. Thus, we are considering parameter uncertainties $p_i \approx 1 + 0.1\delta_i$, $|\delta_i| \le 1$. In other words, we admit parameter variations of 10 percent. Next, we compute $A = -A_1 M$ and

$$B = N_R J^T (JVJ^T)^{-1} = \frac{1}{3} \begin{bmatrix} 2 & 2 & 1 \\ 1 & 1 & 2 \\ 0 & 3 & 0 \\ 1 & -2 & -1 \\ 1 & -2 & -1 \end{bmatrix}$$

$$C = -(JVJ^T)^{-1} JN_L M = \frac{1}{30} \begin{bmatrix} -2 & -1 & -1 & 0 & -1 \\ -1 & -2 & -2 & 3 & -2 \\ -1 & -2 & 1 & 0 & 1 \end{bmatrix}$$

Suppose we wish to eliminate G_4 and g_m. Thus we look for a Gramian of the type $\Sigma = (\sigma_1 I_3, \sigma_4, \sigma_5)$ with $\sigma_4 + \sigma_5$ as small as possible. For simplicity, we additionally require $\sigma_4 = \sigma_5$. Using MATLAB, we find that $\Sigma = \text{diag}(1.3112 I_3, 0.1781 I_2)$ is a balanced Gramian. This suggests deleting the blocks belonging to δ_4 and δ_5, that is, replacing G_4 and g_m by their nominal values. Of course, the symbolic inverse of P can easily be computed in this simplistic example, and we obtain

$$Q = P^{-1} = 1/(G_1 G_2 G_3 + G_1 G_2 G_4 + G_1 G_3 G_4 - g_m G_1 G_2 + G_2 G_3 G_4) \cdot$$

$$\begin{bmatrix} G_2 G_3 + G_2 G_4 + G_3 G_4 - g_m G_2, & G_4(G_2 + G_3), & G_3 G_4 \\ G_2 G_4 + G_3 G_4 - g_m G_2, & (G_1 + G_4)(G_2 + G_3), & (G_1 + G_4)G_3 \\ G_3 G_4, & G_1 G_3 - g_m G_1 + G_3 G_4, & G_1 G_3 + G_1 G_4 - g_m G_1 + G_3 G_4 \end{bmatrix}$$

The symbolic approximate is

$$\hat{Q} = \frac{1}{G_1G_2 + G_1 + G_2} \cdot \begin{bmatrix} 1 + G_2 & 1 + \frac{G_2}{G_3} & 1 \\ 1 & (1+G_1)(1+\frac{G_2}{G_3}) & G_1 + 1 \\ 1 & G_1 - \frac{G_1}{G_3} + 1 & G_1 + 1 \end{bmatrix}$$

and the error bound is $\|Q - \hat{Q}\|_\infty \le 0.7124$. Note that $\|Q\|_\infty \ge 2.2094$ (take $\delta_1 = \ldots = \delta_4 = 0.9$ and $\delta_5 = 1.1$). The relative error bound is not very convincing, but a numerical analysis shows that the true relative error is only 5.41%.

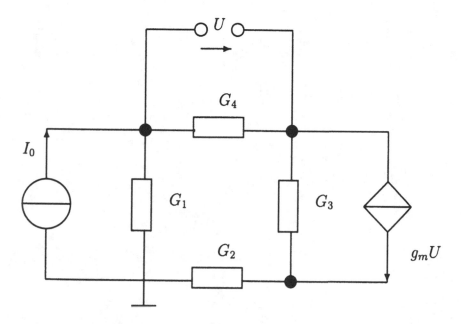

Fig. 7.1. Electrical Network.

Symbolic treatment of the frequency variable: So far, we have assumed P to be evaluated at a certain frequency in order to study the parameter dependency alone. In order to take account of both the frequency and the parameters, consider the Möbius mapping

$$z \mapsto s = \varphi(z) := \frac{1 - z}{1 + z}.$$

This is another linear fractional transformation; it maps the open unit disk to the open right half of the complex plane and vice versa. Now let

$$\tilde{P} \in \mathbb{K}(s; p_1, \ldots, p_r)^{p \times p}, \quad \det \tilde{P} \ne 0,$$

be a non-singular matrix depending rationally on the frequency s and on the parameters $p_1, \ldots, p_r$. Let f be the affine linear transformation

$$f(\delta_1, \ldots, \delta_r) = (p_1, \ldots, p_r), \quad p_i := v_i + m_i \delta_i,$$

then

$$\tilde{P}(s; p_1, \ldots, p_r) = \tilde{P}(\varphi(z); f(\delta_1, \ldots, \delta_r)) =: P(z; \delta_1, \ldots, \delta_r).$$

Define $z =: \delta_0$, then P is a linear fractional transformation of $\delta_0, \ldots, \delta_r$. This signifies that we can treat z like another parameter uncertainty.

Bibliography

1. P. Agathoklis and L. T. Bruton, Practical-BIBO Stability of n-Dimensional Discrete Systems, *IEE Proc.*, **130**, Pt. G, 236–242 (1983).
2. T. Becker and V. Weispfenning, *Gröbner Bases, A Computational Approach to Commutative Algebra*, Graduate Texts in Mathematics **141**, Springer (1993).
3. N. K. Bose, *Applied Multidimensional Systems Theory*, Van Nostrand Reinhold (1982).
4. N. K. Bose and C. Charoenlarpnopparut, Multivariate Matrix Factorization: New Results, *Proceedings of MTNS 98* (1998).
5. S. Boyd, L. El Ghaoui, E. Feron, and V. Balakrishnan, *Linear Matrix Inequalities in System and Control Theory*, SIAM Studies in Applied Mathematics **15** (1994).
6. S. Boyd and C. A. Desoer, Subharmonic Functions and Performance Bounds on Linear Time-Invariant Feedback Systems, *IMA Journal of Mathematical Control and Information* **2**, 153–170 (1985).
7. B. Buchberger, Gröbner Bases: An Algorithmic Method in Polynomial Ideal Theory, in: N. K. Bose (Ed.), *Multidimensional Systems Theory*, Reidel (1985), 184–232.
8. S. M. Carroll, *Lecture Notes on General Relativity*, NSF Report NSF-ITP/97-147 (1997). URL: http://www.itp.ucsb.edu/~carroll/notes/
9. L. Dai, *Singular Control Systems*, Lecture Notes in Control and Information Sciences **118**, Springer (1989).
10. C. A. Desoer and E. S. Kuh, *Basic Circuit Theory*, McGraw-Hill (1969).
11. J. Doyle, Analysis of Feedback Systems with Structured Uncertainties, *IEE Proceedings* **129**, Part D, 242–250 (1982).
12. J. Doyle, A. Packard, and K. Zhou, Review of LFTs, LMIs, and μ, *Proceedings of the 30th Conference on Decision and Control (CDC 91)*, 1227–1232 (1991).
13. D. A. Buchsbaum and D. Eisenbud, What Makes a Complex Exact?, *Journal of Algebra* **25**, 259–268 (1973).
14. D. A. Buchsbaum and D. Eisenbud, Some Structure Theorems for Finite Free Resolutions, *Advances in Mathematics* **12**, 84–139 (1974).
15. W.-K. Chen (Ed.), *The Circuits and Filters Handbook*, CRC Press (1995).
16. Y. Cheng and B. DeMoor, A Multidimensional Realization Algorithm for Parametric Uncertainty Modelling and Multiparameter Margin Problems, *International Journal of Control* **60**, 789–807 (1994).
17. J. C. Cockburn and B. G. Morton, Linear Fractional Representations of Uncertain Systems, *Automatica* **33**, 1263–1271 (1997).
18. D. Eisenbud, *Commutative Algebra with a View Toward Algebraic Geometry*, Graduate Texts in Mathematics **150**, Springer (1995).
19. M. A. Fasoli and F. Pauer, On Certain Linear Spaces of Nilpotent Matrices, *Communications in Algebra* **24**, 3149–3154 (1996).

20. M. A. Fasoli, Finite-Memory Systems, *Multidimensional Systems and Signal Processing* **9**, 291–306 (1998).
21. E. Fornasini and G. Marchesini, State-Space Realization Theory of Two-Dimensional Filters, *IEEE Transactions on Automatic Control* **21**, 484–492 (1976).
22. E. Fornasini and G. Marchesini, Doubly-Indexed Dynamical Systems: State-Space Models and Structural Properties, *Mathematical Systems Theory* **12**, 59–72 (1978).
23. E. Fornasini, P. Rocha, and S. Zampieri, State Space Realization of 2-D Finite-Dimensional Behaviors, *SIAM Journal of Control and Optimization* **31**, 1502–1517 (1993).
24. E. Fornasini and M. E. Valcher, nD Polynomial Matrices with Applications to Multidimensional Signal Analysis, *Multidimensional Systems and Signal Processing* **8**, 387–408 (1997).
25. P. A. Fuhrmann, On Strict System Equivalence and Similarity, *International Journal of Control* **25**, p. 5–10, 1977.
26. K. Gałkowski, Transformation of the Transfer Function Variables of the Singular n-Dimensional Roesser Model, *International Journal of Circuit Theory and Applications* **20**, 63–74 (1992).
27. F. R. Gantmacher, *The Theory of Matrices*, reprint, AMS Chelsea Publ. (1998).
28. G. M. Greuel, G. Pfister, and H. Schönemann, Singular Reference Manual, in: *Reports on Computer Algebra* **20**, Centre for Computer Algebra, University of Kaiserslautern (1998). URL: http://www.mathematik.uni-kl.de/~zca/Singular
29. J. P. Guiver and N. K. Bose, Polynomial Matrix Primitive Factorization Over Arbitrary Coefficient Fields and Related Results, *IEEE Transactions on Circuits and Systems* **29**, 649–657 (1982).
30. J. P. Guiver and N. K. Bose, Causal and Weakly Causal 2-D Filters with Applications in Stabilization, in: N. K. Bose (Ed.), *Multidimensional Systems Theory*, Reidel (1985), 52–100.
31. W. W. Hager, Updating the Inverse of a Matrix, *SIAM Review* **31**, 221–239 (1989).
32. D. Hinrichsen and D. Prätzel-Wolters, Solution Modules and System Equivalence, *International Journal of Control* **32**, 777–802, 1980.
33. T. Kaczorek, *Two-Dimensional Linear Systems*, Lecture Notes in Control and Information Science **68**, Springer (1985).
34. T. Kaczorek, Singular Multidimensional Roesser Model, *Bulletin of the Polish Academy of Sciences*, Tech. Sci. **36**, 327–335 (1988).
35. T. Kailath, *Linear Systems*, Prentice-Hall (1980).
36. M. Kuijper, *First-order Representations of Linear Systems*, Birkhäuser (1994).
37. N. M. Kumar, On Two Conjectures about Polynomial Rings, *Inventiones Mathematicae* **46**, 225-236 (1978).
38. S.-Y. Kung, B. C. Lévy, M. Morf, and T. Kailath, New Results in 2-D Systems Theory, Part II: 2-D State-Space Models — Realization and the Notions of Controllability, Observability, and Minimality, *Proceedings of the IEEE* **65**, 945–961 (1977).
39. E. Kunz, *Introduction to Commutative Algebra and Algebraic Geometry*, Birkhäuser (1985).
40. T. Y. Lam, *Serre's Conjecture*, Lecture Notes in Mathematics **635**, Springer (1978).
41. Z. Lin, On Matrix Fraction Descriptions of Multivariable Linear n-D Systems, *IEEE Transactions on Circuits and Systems* **35**, 1317–1322 (1988).
42. Z. Lin, Feedback Stabilizability of MIMO n-D Linear Systems, *Multidimensional Syst. Signal Process.*, **9**, 149–172 (1998).

43. W. M. Lu, K. Zhou, and J. C. Doyle, Stabilization of LFT Systems, *Proceedings of the 30th Conference on Decision and Control (CDC 91)*, 1239–1244 (1991).

44. S. Mandal, Number of Generators for Modules over Laurent Polynomial Rings, *Journal of Algebra* **80**, 306–313 (1983).

45. S. A. Miri and J. D. Aplevich, Modeling and Realization of n-Dimensional Linear Discrete Systems, *Multidimensional Systems and Signal Processing* **9**, 241–253 (1998).

46. M. Morf, B. C. Lévy, and S.-Y. Kung, New Results in 2-D Systems Theory, Part I: 2-D Polynomial Matrices, Factorization, and Coprimeness, *Proceedings of the IEEE* **65**, 861–872 (1977).

47. D. G. Northcott, *Finite Free Resolutions*, Cambridge Tracts in Mathematics **71**, Cambridge University Press (1976).

48. U. Oberst, Multidimensional Constant Linear Systems, *Acta Applicandae Mathematicae* **20**, 1–175 (1990).

49. U. Oberst, Variations on the Fundamental Principle for Linear Systems of Partial Differential and Difference Equations with Constant Coefficients, *Applicable Algebra in Engineering, Communication and Computing* **6**, 211–243 (1995).

50. A. Packard and J. Doyle, The Complex Structured Singular Value, *Automatica* **29**, 71–109 (1993).

51. V. P. Palamodov, *Linear Differential Operators with Constant Coefficients*, Grundlehren der mathematischen Wissenschaften **168**, Springer (1970).

52. L. Pernebo, Notes on Strict System Equivalence, *International Journal of Control* **25**, 21–38 (1977).

53. H. Pillai and S. Shankar, A Behavioral Approach to Control of Distributed Systems, *SIAM Journal on Control and Optimization* **37**, 388–408 (1998).

54. J. F. Pommaret, Dualité Différentielle et Applications, *Comptes Rendus de l'Académie des Sciences de Paris, Série I*, **320**, 1225–1230 (1995).

55. J. F. Pommaret and A. Quadrat, Generalized Bézout Identity, *Applicable Algebra in Engineering, Communication and Computing* **9**, 91–116 (1998).

56. J. F. Pommaret, Einstein Equations Do Not Admit a Generic Potential, *Proceedings of the 6th International Conference on Differential Geometry and Applications*, Brno, Czech Republic (1995).

57. A. C. Pugh, J. S. McInerney, M. S. Boudellioua, D. S. Johnson and G. E. Hayton, A Transformation for 2-D Linear Systems and a Generalisation of a Theorem of Rosenbrock, *International Journal of Control* **71**, 491–503 (1998)

58. P. Rocha and J. C. Willems, State for 2-D Systems, *Linear Algebra and its Applications* **122–124**, 1003–1038 (1989).

59. P. Rocha, *Structure and Representation of 2D Systems*, Ph. D. Thesis, University of Groningen, The Netherlands (1990).

60. P. Rocha and J. C. Willems, On the Representation of 2D Systems, *Kybernetica* **27**, 225–230 (1991).

61. P. Rocha and J. C. Willems, Controllability of 2-D Systems, *IEEE Transactions on Automatic Control* **36**, 413–423 (1991).

62. R. P. Roesser, A Discrete State-Space Model for Linear Image Processing, *IEEE Transactions on Automatic Control* **20**, 1–10 (1975).

63. H. H. Rosenbrock, *State-Space and Multivariable Theory*, Nelson (1970).

64. J. Rosenthal, J. M. Schumacher, and E. V. York, On Behaviors and Convolutional Codes, *IEEE Transactions on Information Theory* **42**, 1881–1891 (1996).

65. W. Rupprecht, *Netzwerksynthese*, Springer (1972).

66. M. G. Safonov, Stability Margins of Diagonally Perturbed Multivariable Feedback Systems, *IEE Proceedings* **129**, Part D, 251–256 (1982).

67. S. Shankar and V. R. Sule, Algebraic Geometric Aspects of Feedback Stabilization, *SIAM J. Cont. Opt.*, **30**, 11–30 (1992).

68. P. Slepian, *Mathematical Foundations of Network Analysis*, Springer Tracts in Natural Philosophy **16** (1968).
69. R. Sommer, E. Hennig, G. Dröge, and E. H. Horneber, Equation-Based Symbolic Approximation by Matrix Reduction with Quantitative Error Prediction, *Alta Frequenza – Rivista di Elettronica* **6** (1993).
70. L. Staiger, Subspaces of $GF(q)^{\omega}$ and Convolutional Codes, *Information and Control* **59**, 148–183 (1983).
71. J. Stoer and C. Witzgall, Transformations by Diagonal Matrices in a Normed Space, *Numerische Mathematik* **4**, 158–171 (1962).
72. T. Sugie and M. Kawanishi, μ Analysis/Synthesis Based on Exact Expression of Physical Parameter Variations and its Application, *Proceedings of the 3rd European Control Conference (ECC 95)*, 159–164 (1995).
73. V. R. Sule, Feedback Stabilization over Commutative Rings: The Matrix Case, *SIAM J. Cont. Opt.*, **32**, 1675–1695 (1994); corrigendum ibid., **36**, 2194–2195 (1998).
74. M. N. S. Swamy and K. Thulasiraman, *Graphs, Networks, and Algorithms*, Wiley (1981).
75. M. E. Valcher, On the Decomposition of Two-Dimensional Behaviors, *Multidimensional Systems and Signal Processing*, **11**, 49–65 (2000).
76. M. Vidyasagar, *Control System Synthesis: A Factorization Approach*, MIT Press (1985).
77. W. Wang, J. Doyle, C. Beck and K. Glover, Model Reduction of LFT Systems, *Proceedings of the 30th IEEE Conference on Decision and Control (CDC 91)*, 1233–1238, (1991).
78. S. Weiland, *Theory of Approximation and Disturbance Attenuation for Linear Systems*, Ph. D. Thesis, University of Groningen, The Netherlands (1991).
79. J. C. Willems, From Time Series to Linear System—Part I. Finite Dimensional Linear Time Invariant Systems, *Automatica* **22**, 561–580 (1986).
80. J. C. Willems, Models for Dynamics, *Dynamics Reported* **2**, 171–269 (1989).
81. J. C. Willems, Paradigms and Puzzles in the Theory of Dynamical Systems, *IEEE Transactions on Automatic Control* **36**, 259–294 (1991).
82. J. Wood, E. Rogers, and D. H. Owens, Further Results on the Stabilization of Multidimensional Systems, *Proc. 4th European Control Conference* (ECC 97), Vol. 4, Brussels (1997).
83. J. Wood, E. Rogers and D. H. Owens, A Formal Theory of Matrix Primeness, *Mathematics of Control, Systems and Signals* **11**, 40–78 (1998).
84. J. Wood, E. Rogers, and D. H. Owens, Controllable and Autonomous nD Systems, *Multidimensional Systems and Signal Processing* **10**, 33–70 (1999).
85. J. Wood and E. Zerz, Notes on the Definition of Behavioral Controllability, *Systems and Control Letters* **37**, 31–37 (1999).
86. L. Xu, O. Saito, and K. Abe, The Design of Practically Stable nD Feedback Systems, *Automatica*, **30**, 1389–1397 (1994).
87. D. C. Youla, H. A. Jabr, and J. J. Bongiorno, Modern Wiener-Hopf design of optimal controllers, *IEEE Trans. Aut. Cont.*, **21**, 3–13 and 319–338 (1976).
88. D. C. Youla and G. Gnavi, Notes on n-Dimensional System Theory, *IEEE Transactions on Circuits and Systems* **26**, 105–111 (1979).
89. D. C. Youla and P. F. Pickel, The Quillen–Suslin Theorem and the Structure of n-Dimensional Elementary Polynomial Matrices, *IEEE Transactions on Circuits and Systems* **31**, 513– 518 (1984).
90. E. Zerz, Primeness of Multivariate Polynomial Matrices, *Systems and Control Letters* **29**, 139–145 (1996).
91. E. Zerz, Primeness of Multivariate Laurent Polynomial Matrices, *Proceedings of the 4th European Control Conference (ECC 97)* **2** (1997).

92. E. Zerz, LFT Representations of Parametrized Polynomial Systems, *IEEE Transactions on Circuits and Systems I* **46**, 410–416 (1999).

93. E. Zerz, Linear Fractional Representations of Polynomially Parametrized Descriptor Systems, *Proceedings of the 3rd Portuguese Conference on Automatic Control (CONTROLO 98)*, 157–162 (1998).

94. E. Zerz, Coprime Factorizations of Multivariate Rational Matrices, to appear in *Mathematics of Control, Signals and Systems.*

95. E. Zerz, On Strict System Equivalence for Multidimensional Systems, accepted for publication in the *International Journal of Control*, 1999.

96. E. Zerz, First-order Representations of Discrete Linear Multidimensional Systems, accepted for publication in *Multidimensional Systems and Signal Processing.*

97. K. Zhou, J. C. Doyle, and K. Glover, *Robust and Optimal Control*, Prentice-Hall (1996).

Index

Lecture Notes in Control and Information Sciences

Edited by M. Thoma

1997–2000 Published Titles: